Advanced
Modular
Mathematics

PURE
MATHEMATICS 1

Stephen Webb

**SECOND
EDITION**

NATIONAL
EXTENSION
COLLEGE

Unit P1

Published by HarperCollins Publishers Limited
77–85 Fulham Palace Road
Hammersmith
London W6 8JB

www.CollinsEducation.com
On-line Support for Schools and Colleges

This book was written by Stephen Webb for the National Extension College Trust Ltd.

British Library Cataloguing in Publication Data
A catalogue record for this publication is available from the British Library.

Original internal design: Derek Lee
Cover design and implementation: Terry Bambrook
Page layout: Mary Bishop and Janet Oswell
Project Editors: Hugh Hillyard-Parker and Margaret Levin
Printed and bound: Martins the Printers Ltd., Berwick-upon-Tweed

The authors and publishers thank Dave Wilkins for his comments on this book.

The National Extension College is an educational trust and a registered charity
with a distinguished body of trustees. It is an independent, self-financing organisation.

Since it was established in 1963, NEC has pioneered the development of flexible
learning for adults. NEC is actively developing innovative materials and systems for
distance-learning options from basic skills and general education to degree and
professional training.

For further details of NEC resources that support Advanced Modular Mathematics,
and other NEC courses, contact NEC Customer Services:

National Extension College Trust Ltd
18 Brooklands Avenue
Cambridge CB2 2HN
Telephone 01223 316644, Fax 01223 313586

You might also like to visit:

www.fireandwater.com
The book lover's website

Contents

P1

Advanced Modular Mathematics

Permissions

We are grateful to the following examination boards for permission to reproduce in the Summary exercises the following questions from past examination papers.

The examination boards credited below accept no responsibility whatsoever for the accuracy or method of working in the answers given in the Solutions section, which are entirely the responsibility of the author.

Edexcel

(including questions published by the London Board)

Section 1: Exercises 2, 3, 5, 7

Section 3: Exercises 1, 2, 8, 9, 10, 11

Section 4: Exercises 3, 4, 5, 7, 8, 9

Section 5: Exercises 4, 7, 8, 11, 12

Section 7: Exercises 12, 13, 15

Section 8: Exercises 4, 5, 6, 8, 9

OCR

(including questions published by the Oxford and Cambridge Boards)

Section 1: Exercises 6, 8, 9, 10

Section 2: Exercises 6 , 7, 8, 9, 10, 11, 12, 13, 14

Section 3: Exercises 5, 11, 13

Section 4: Exercise 10

Section 5: Exercises 1, 2, 3, 10

Section 7: Exercises 8, 9, 11

Section 8: Exercises 1, 2, 7

Assessment and Qualifications Alliance

(including questions published by the Associated Examining Board)

Section 1: Exercises 1, 4, 12

Section 2: Exercise 5

Section 3: Exercises 3, 6, 7, 12

Section 4: Exercises 1, 2, 6

Section 5: Exercises 5, 6, 9

Section 7: Exercises 6, 7, 14

Section 8: Exercise 3

SECTION

1

Algebra

INTRODUCTION This is the fundamental skill that you really have to acquire if your study of 'A' level mathematics is to be successful. There is nothing very difficult in the ideas, but you do have to keep up a high level of accuracy in completing fairly routine tasks. This can only come with plenty of practice, so there is a large number of questions for you to try in this section, and, of course, answers in the back so that you can check your working is correct.

The basic techniques include expanding brackets and the reverse of this, factorising: this is what we look at first, followed by methods for more advanced factorisation. We then look in more detail at one particular family of functions, the quadratics – their graphs, equations and inequalities – and finally methods for working out where a line and a curve cross.

Polynomials

OCR P2 5.2.3 (a)

A polynomial is a collection of terms involving a variable, such as $x^3 - 2x^2 + 5$ or simply $3x + 1$. When the highest power of the variable is 2, for example, $x^2 - 3x + 7$, it is called a *quadratic* polynomial, or quadratic for short. A *cubic* polynomial, like $x^3 - 4x^2 + 5x + 4$, has a highest power of 3. It can also be called a polynomial of *order* 3. As you can see, polynomials are usually written with the term involving the highest power first, then the powers go down until the term without the variable called the *constant* term.

We can label polynomials by giving them a letter and writing $f(x)$ or $g(x)$ if our variable is x, so that we could write our cubic polynomial above as:

$$f(x) = x^3 - 4x^2 + 5x + 4$$

The variable is not always x – it can sometimes be another letter, t say, in which case our cubic would be written as:

$$f(t) = t^3 - 4t^2 + 5t + 4$$

Once we've labelled the polynomial like this, we can simply write $f(x)$ instead of writing it down in full when we need to refer to it. Now we can take another polynomial, our quadratic above for example, and call it $g(x)$, so that:

$$g(x) = x^2 - 3x + 7$$

Adding and subtracting polynomials

If we want to add the polynomials, $f(x) + g(x)$, we add the like terms together, so that the x^2 terms are added together, and so on. For the moment, we can set this sum out as though we were adding numbers, although normally you would write them on one line:

$$\begin{array}{ll}
\text{f}(x) & x^3 - 4x^2 + 5x \;\; + 4 \\
+\,\text{g}(x) & \underline{\hspace{0.7cm} x^2 \;- 3x \;\; + 7} \\
& x^3 - 3x^2 + 2x \;\; + 11
\end{array}$$

We use the same method, of combining like terms, when it comes to subtracting polynomials, but we have to be *very* careful to watch the signs. Probably the best way until you're very confident is to take the subtraction in stages. Suppose you want to subtract g(x) from f(x). You could write:

$$\text{f}(x) - \text{g}(x) = (x^3 - 4x^2 + 5x + 4) - (x^2 - 3x + 7)$$

Since there is nothing outside the first bracket on the right-hand side we don't need it and we can write the polynomial without it. The last bracket has a negative sign outside and this means that *all* the signs have to be changed if we want to remove it. When this has been done, we add the like terms together as we did when we added the polynomials:

$$\begin{aligned}
\text{f}(x) - \text{g}(x) &= x^3 - 4x^2 + 5x + 4 - x^2 + 3x - 7 \\
&= x^3 - 5x^2 + 8x - 3
\end{aligned}$$

Multiplying polynomials

You probably already know that to multiply out a bracket like $2x(3x-4)$, you multiply both terms inside the bracket by the $2x$, to give

$$2x(3x-4) = 6x^2 - 8x$$

To expand an expression like $(2x+3)(3x-4)$ we split the first bracket into two parts and multiply out each separately, collecting together any like terms:

$$\begin{aligned}
(2x+3)(3x-4) &= 2x(3x-4) + 3(3x-4) \\
&= 6x^2 - 8x + 9x - 12 \\
&= 6x^2 + x - 12
\end{aligned}$$

When multiplying more complicated polynomials, you need to be quite systematic. Suppose we had to find the product of two quadratics, for example:

$$(1 - 3x + x^2)(1 - 4x - x^2)$$

One way would be to multiply everything in the right-hand bracket by each of the terms in the left-hand bracket in turn, and then collect like terms together:

$$(1 - 3x + x^2)(1 - 4x - x^2) = 1 - 4x - x^2 \quad -3x + 12x^2 + 3x^3 \quad +x^2 - 4x^3 - x^4$$

$$= 1 - 7x + 12x^2 - x^3 - x^4$$

Practice questions A

1 Expand:
  (a) $3x(2x+4)$
  (b) $3(2x-1)$
  (c) $-4(2x-3)$
  (d) $-2x(1+x)$
  (e) $3x(1-3x)$
  (f) $-4x(5-3x)$

2 Expand:
  (a) $(4x+1)(2x-3)$
  (b) $(2x-1)(3x-2)$
  (c) $(2x-3)(2x+3)$
  (d) $(1-x)(1+2x)$
  (e) $(1+3x)(2-x)$
  (f) $(1+4x)(3x-2)$

3 Expand:
  (a) $3x(1-2x+x^2)$
  (b) $(1-2x)(1+x+2x^2)$
  (c) $(1-x)(1+x+x^2)$
  (d) $(1-2x+x^2)(1+2x-x^2)$
  (e) $(1-x-x^2)(3+2x+3x^2)$
  (f) $(1+x+x^2)^2$

Identities

The equation $(x-1)^2 = 4$ has only two solutions for x, given by $x - 1 = \pm 2 \Rightarrow$ $x = 3$ or $x = -1$. On the other hand, an equation like $(x-1)^2 = x^2 - 2x + 1$ is always true, no matter what values of x we put in. In this case, we call it an *identity* and to distinguish this from an equation, we write it:

$$(x-1)^2 \equiv x^2 - 2x + 1$$

with an extra line in the equal sign.

Sometimes we are given an equation with some unknown coefficients and asked to find the value of the coefficients that would make the equation true for all values of x, i.e. make it an identity. Suppose, for example, that we are given

$$(x+1)(x^2 + ax + b) \equiv x^3 + 3x^2 + 5x + 3$$

What values of a and b must we choose to make the two sides equal for any value of x? There are two approaches. Either we substitute suitable values of x, probably ending up with simultaneous equations, or we multiply out the left hand side and compare the coefficients on the two sides. Let's try each of these methods.

Values

If we take $x = 1$ and put this value into both sides of the identity, we have:

$$2(1 + a + b) = 1 + 3 + 5 + 3 \qquad \Rightarrow 2 + 2a + 2b = 12$$
$$\Rightarrow 2a + 2b = 10$$
$$\Rightarrow a + b = 5 \qquad \dots \text{①}$$

Then taking $x = 2$,

$$3(4 + 2a + b) = 8 + 12 + 10 + 3 \quad \Rightarrow \quad 12 + 6a + 3b = 33$$
$$\Rightarrow 6a + 3b = 21$$
$$\Rightarrow 2a + b = 7 \qquad \dots \text{②}$$

Subtracting ① from ② gives $a = 2 \Rightarrow b = 3$

Coefficients

Multiplying out the brackets gives:

$$x^3 + ax^2 + bx + x^2 + ax + b \quad = x^3 + 3x^2 + 5x + 3$$
$$x^3 + (a+1)x^2 + (a+b)x + b \ = x^3 + 3x^2 + 5x + 3$$

If we want these to be equal for any value of x, they must be exactly the same. Equating the x^2 coefficients:

$$a + 1 = 3 \Rightarrow a = 2$$

Equating the constant term:

$$b = 3,$$

which are the solutions we found before.

Here is another example, which involves completing the square:

| **Example** | Find the numbers m and n such that |

$$5 + 4x - x^2 = m - (x - n)^2$$

for all real values of x.

Solution	$m - (x - n)^2$	$= m - (x^2 - 2nx + n^2)$
		$= m - n^2 + 2nx - x^2$

Comparing coefficients, $5 = m - n^2$ ... ①

and $4 = 2n$ ... ②

From ②, $n = 2$; putting this into ① gives $5 = m - 4 \Rightarrow m = 9$.

Practice questions B

1   Find the values of the constants in the following identities

  (a) $A(x^2 - 1) + B(x - 1) + C \equiv (3x - 1)(x + 1)$

  (b) $x^4 + \dfrac{4}{x^4} = \left(x^2 - \dfrac{A}{x^2}\right)^2 + B$

  (c) $x^4 + Ax^3 + 5x^2 + x + 3$
$\equiv (x^2 + 4)(x^2 - x + B) + Cx + D$

  (d) $x^4 + x^2 + x + 1 \equiv (x^2 + A)(x^2 - 1) + Bx + C$

  (e) $2x^3 + 3x^2 - 14x - 5 \equiv (Ax + B)(x + 3)(x + 1) + C$

  (f) $x^3 + Ax^2 - 23x + B \equiv (x - C)(x - 4)(x + 5)$

Factorising

OCR P2 5.2.3 (d)

This is the reverse procedure to expanding. At the simplest, there is some common factor between the terms: $2x^2 - 4x$ has the common factor of $2x$ for example. This factor is placed outside a bracket – inside the bracket are the terms which when multiplied by the $2x$ give the original expression:

$$2x^2 - 4x = 2x(x - 2)$$

We try and find the *highest* common factor: the biggest number and the highest power of the variable(s) that divide all the terms, so for example:

$$12x^2y^2 - 8xy^3 = 4xy^2(3x - 2y)$$

Factorising quadratics

Suppose we have to factorise $x^2 + 7x + 12$, which means expressing it in the form $(x + a)(x + b)$ where a and b are numbers we have to find. If we multiply these brackets out, giving $x^2 + (a + b)x + ab$ and compare the quadratics, we can see that we want to choose a and b so that $ab = 12$ and $a + b = 7$. This is quite easy: one must be 3, the other 4 and the factors are $(x + 3)(x + 4)$. Note that if the original expression had been $x^2 - 7x + 12$, we would still have $ab = 12$ but now $a + b = -7$. Since the product ab is positive, the numbers a and b have the same signs; as the sum $a + b$ is negative, they must both be negative and so the factors in this case would be $(x - 3)(x - 4)$.

If ab is negative, a and b have opposite signs. Then we have to look at the sum $a + b$, the x-coefficient. If this is positive, the larger of a and b is positive: if negative, the larger is negative. So for example with $x^2 + 2x - 8$: $ab = -8$ means that a and b have opposite signs. $a + b = 2$, so the larger is positive: $+4, -2$ with factors $(x + 4)(x - 2)$. With $x^2 - x - 12$: $ab = -12$, again opposite signs but this time the larger is negative: $-4, +3$ with factors $(x - 4)(x + 3)$.

We have to use a slightly different method when the coefficient of x^2 is not 1. Maybe you already have a way of doing this which works for you – fine!

Just make sure that you can do the practice examples without too much trouble. Otherwise, here is one way – there are many others.

Suppose we have to factorise a quadratic like $6x^2 - 5x - 6$. We proceed in a series of steps.

1 Multiply the x^2-coefficient by the constant. This gives $6 \times -6 = -36$.

2 Find two numbers whose product is the number just found (i.e. -36) and whose sum is the x-coefficient (in this case -5). The numbers are -9 and $+4$.

3 Rewrite the original quadratic with the x-term replaced by the sum of two x-terms whose coefficients are the numbers found in Step 2. So instead of $-5x$ we put $-9x + 4x$ and $6x^2 - 5x - 6$ becomes $6x^2 - 9x + 4x - 6$.

4 Take common factors out of the first two terms and then the next two terms

$$6x^2 - 9x + 4x - 6 = 3x(2x - 3) + 2(2x - 3)$$

The brackets should be the same. (You may have to adjust the sign.)

5 Combine the two into a product of two brackets

$$3x(2x - 3) + 2(2x - 3) = (3x + 2)(2x - 3)$$

Have a look at another example and then try some yourself.

Example	Factorise: $12x^2 + 11x - 5$

Solution	$12 \times -5 = -60$ so we want two numbers, product -60 and sum 11. These are $+15$ and -4. Rewrite the $11x$ as $15x - 4x$ and we have:

$$12x^2 + 15x - 4x - 5 = 3x(4x + 5) - 1(4x + 5)$$
$$= (3x - 1)(4x + 5)$$

Note the adjustment of sign to make the last bracket the same as the first.

Have a go at some of these.

Practice questions C

1 Factorise:

(a) $3x - 6x^2$

(b) $4xy + 8y$

(c) $pq^3 - p^2q$

(d) $12x^2 + 16x^3$

(e) $6p - 9q$

(f) $x + 5x^2$

2 Factorise:

(a) $x^2 + 3x + 2$

(b) $x^2 + 6x + 8$

(c) $x^2 - 7x + 12$

(d) $x^2 - 5x + 6$

(e) $x^2 - 2x - 8$

(f) $x^2 + 3x - 10$

(g) $x^2 - 4x - 21$

(h) $x^2 - 10x + 25$

(i) $x^2 + 6x + 9$

(j) $x^2 + x - 20$

3 Factorise:

(a) $2x^2 + 3x - 2$

(b) $6x^2 + 5x + 1$

(c) $8x^2 - 2x - 3$

(d) $3x^2 + 7x - 6$

(e) $6x^2 + 11x - 2$

(f) $6x^2 - 5x - 4$

Cubic polynomials

OCR P2 5.2.3 (d)

There is no equivalent method for factorising polynomials of order 3 or more and we have to use a different approach. This involves finding *one* factor first of all, which allows us to split the original polynomial into two parts. We can see the process at work if we take a numerical example. Suppose we had to find the prime factors of 715. We can see immediately that one of the factors must be 5, since the number ends with a 5. By dividing the number by the factor we have discovered, we see that $715 = 5 \times 143$. We now have a smaller number, 143, to factorise and so on.

Returning to the problem of factorising a polynomial, how do we find one of its factors to begin with? To see how to do this, let's start the other way round. Suppose we know the factors already, say $f(x) = (x + 1)(x - 3)(2x - 5)$. When we substitute the values that make each of the brackets zero in turn, we make the whole polynomial zero since the factors are all multiplied together:

when

$$(x + 1) = 0, \quad x = -1 \quad : \quad f(-1) = 0 \times (-1 - 3) \times (-2 - 5) = 0$$
$$(x - 3) = 0, \quad x = 3 \quad : \quad f(3) = (3 + 1) \times 0 \times (6 - 5) = 0$$
$$(2x - 5) = 0, \quad x = \frac{5}{2} \quad : \quad f\left(\frac{5}{2}\right) = \left(\frac{5}{2} + 1\right) \times \left(\frac{5}{2} - 3\right) \times 0 = 0$$

Luckily enough, this works in reverse so that for example if we know that $f(-1) = 0$ then $(x + 1)$ is a factor, and if $f(3) = 0$ we can say that $(x - 3)$ is a factor.

In general for any polynomial $f(x)$

$$f(a) = 0 \Leftrightarrow (x - a) \text{ is a factor of } f(x).$$

This is called the *Factor theorem* .

So to find a factor of a polynomial, we try values of x until we find one for which $f(x) = 0$, say when $x = a$ i.e. $f(a) = 0$. Then we can say that $(x - a)$ is a factor.

Example	Find the factors of

$$f(x) = x^3 - 2x^2 - 5x + 6$$

Solution	We only need to try the factors of the constant term, here 6, so we don't need for example to try $x = 4$ or $x = 5$.

$$f(1) = 1 - 2 - 5 + 6 = 0 \quad \Rightarrow \quad (x - 1) \text{ is a factor}$$
$$f(-1) = -1 - 2 + 5 + 6 = 8 \neq 0 \quad \Rightarrow \quad (x + 1) \text{ is not a factor}$$
$$f(2) = 8 - 8 - 10 + 6 = -4 \neq 0 \quad \Rightarrow \quad (x - 2) \text{ is not a factor}$$
$$f(-2) = -8 - 8 + 10 + 6 = 0 \quad \Rightarrow \quad (x + 2) \text{ is a factor}$$
$$f(3) = 27 - 18 - 15 + 6 = 0 \quad \Rightarrow \quad (x - 3) \text{ is a factor}$$

We can stop there, since a cubic can only have three different linear factors at most. We have found that $f(x)$ has the three factors $(x - 1)$, $(x + 2)$ and $(x - 3)$.

Knowing one or more factors, we can deduce equations which allow us to find the value of unknown coefficients.

Example	Find the value of the constant a, given that $(x - 2)$ is a factor of

$$f(x) = 3x^3 - 2x^2 + ax - 8$$

| **Solution** | If $(x - 2)$ is a factor, |

$$f(2) = 0 \Rightarrow \quad 24 - 8 + 2a - 8 \quad = 0$$
$$8 + 2a \quad = 0 \Rightarrow a = -4$$

Practice questions D

1   Show that:

  (a) $f(x) = x^3 - 3x^2 - 4x + 12$ has the factors
        $(x - 2)$, $(x + 2)$ and $(x - 3)$

  (b) $f(x) = x^3 - 7x + 6$ has the factors
        $(x - 1)$, $(x - 2)$ and $(x + 3)$

  (c) $f(x) = x^3 - 3x^2 - 6x + 8$ has the factors
        $(x - 1)$, $(x + 2)$ and $(x - 4)$

2   Find the factors for:

  (a) $f(x) = x^3 - 3x^2 + 4$

  (b) $f(x) = x^3 - 9x^2 + 26x - 24$

  (You need only try values for $x \leq 4$)

3   Find the value of the constant a for which the
        polynomial $f(x) = x^3 + 4x^2 - 10x + a$
        has a factor:

  (a) $x - 1$ (b) $x + 2$ (c) x

Quite frequently we are only able to find one linear factor using the method of substituting values. We then have a choice of methods for finding the other factor (which will be quadratic if the original polynomial was cubic).
One quick method is to treat the polynomial and its factors as an identity.
Let's have a look at an example of this.

| **Example** | Show that the polynomial |

$$f(x) = x^3 - x^2 - 3x - 9$$

has a factor $(x - 3)$ and find the remaining factor.

| **Solution** | To show that $f(x)$ has a factor $(x - 3)$, we have to show that $f(3) = 0$. |

$$f(3) = 27 - 9 - 9 - 9 = 0 \quad \Rightarrow \quad (x - 3) \text{ is a factor}$$

Suppose the other (quadratic) factor is $ax^2 + bx + c$. We then try and solve the identity

$$f(x) = x^3 - x^2 - 3x - 9 \equiv (x - 3)(ax^2 + bx + c)$$

comparing x^3-coefficients, $1 = a \quad \Rightarrow \quad a = 1$

comparing constants, $-9 = -3c \Rightarrow c = 3$

That was the easy bit. We now have to look at either the x^2-coefficient or the x-coefficient. On the RHS, the x^2-terms will be the sum of $x \times bx$ and $-3 \times ax^2$, so the x^2-coefficient is $b - 3a$. On the LHS, the x^2-coefficient is -1, so we have

$$b - 3a = -1 \Rightarrow b = 3a - 1 = 2 \quad \text{since } a = 1$$

So the other factor is $x^2 + 2x + 3$ (which doesn't factorise any further),

so $f(x) = (x - 3)(x^2 + 2x + 3)$

Have a look at one more example of this.

| **Example** | Find a, b and c so that |

$$4x^3 - 12x^2 - 19x + 12 \equiv (2x - 1)(ax^2 + bx + c)$$

Solution	x^3-coeff	$4 = 2a$	$\Rightarrow a = 2$
	constant	$12 = -c$	$\Rightarrow c = -12$
	x^2-coeff	$-12 = -a + 2b$	$\Rightarrow b = -5$

i.e. the quadratic factor is $2x^2 - 5x - 12$.

In practice, it's quite easy to see what a and c are, just by comparing the first and last terms so the only problem is finding b. For example, if we know that $(3x - 1)$ is a factor of $6x^3 + x^2 + 2x - 1$ we could write down the quadratic straight away as $2x^2 + bx + 1$, having found a to be 2 and c to be 1 in our head. Then from the x^2-coefficient, $1 = 3b - 2 \Rightarrow b = 1$ and the quadratic factor is $2x^2 + x + 1$.

Practice questions E

1 Find values of a, b and c for which:
  (a) $3x^3 - 5x^2 + 12x - 10 \equiv$
  $(x - 1)\,(ax^2 + bx + c)$
  (b) $x^3 - 3x^2 + 8x - 12 \equiv (x - 2)\,(ax^2 + bx + c)$
  (c) $6x^3 + 7x^2 + 7x + 15 \equiv$
  $(2x + 3)\,(ax^2 + bx + c)$
  (d) $x^3 - 9x - 10 \equiv (x + 2)\,(ax^2 + bx + c)$

2 Find the quadratic factor of f(x) when:
  (a) f(x) $= 2x^3 - 5x^2 + 2x - 15$ has factor $(x - 3)$
  (b) f(x) $= 3x^3 + 3x^2 + x + 14$ has factor $(x + 2)$
  (c) f(x) $= 6x^3 + 11x^2 - 2$ has factor $(2x + 1)$
  (d) f(x) $= 8x^3 + 4x - 3$ has factor $(2x - 1)$

3 Show that f(x) $= x^3 - 4x^2 + 3$ has a factor $(x - 1)$ and find the other, quadratic, factor.

4 Show that f(x) $= 2x^3 - 3x^2 - 8x - 3$ has a factor $(x + 1)$. Find the other quadratic factor, factorise this and hence express f(x) as the product of three linear factors.

5 Factorise the following expressions into three linear factors:
  (a) f(x) $= x^3 - 2x^2 - 5x + 6$
  (b) f(x) $= 3x^3 - 4x^2 - 5x + 2$
  (c) f(x) $= x^3 - 12x + 16$
  (d) f(x) $= 2x^3 + 5x^2 + x - 2$
  (e) f(x) $= 2x^3 + 7x^2 - 7x - 12$
  (f) f(x) $= 2x^3 - 7x^2 + 9$
  (g) f(x) $= x^3 - x^2 - 6x$ [take out the factor of x first]
  (h) f(x) $= x^3 + x^2 - 8x - 12$
  (i) f(x) $= 2x^3 - 9x^2 - 6x + 5$
  (j) f(x) $= x^3 - 3x^2 + 4$

Dividing algebraically follows much the same pattern as arithmetic 'long division'. Let's have a look at an example of this:

$$\frac{x^3 - 3x^2 + 2x + 7}{x + 1}$$

We'll set it out as we would for arithmetic division:

$$x + 1 \;\overline{\smash{\big)}\, x^3 - 3x^2 + 2x + 7}$$

The difference is that we only worry about the highest power of x in the numerator being divided and the highest power of x in the denominator doing the dividing, in the immediate case x^3 and x respectively. This gives x^2; now we proceed exactly as before:

$$
\begin{array}{r}
x^2 \\
x + 1 \,\overline{\smash{\big)}\, x^3 - 3x^2 + 2x + 7} \\
\underline{x^3 + x^2 } \\
-4x^2
\end{array}
$$

Remember that we *subtract* the $(x^3 + x^2)$ from $(x^3 - 3x^2)$. We can look on this as reversing the signs of $x^3 + x^2$, giving $-x^3 - x^2$, and *adding* this to $x^3 - 3x^2$, giving $-4x^2$. As we have already we bring the next term down and continue:

$$
\begin{array}{r}
x^2 - 4x\ + 6 \\
x + 1\ \overline{\big)\ x^3 - 3x^2 + 2x + 7} \\
\underline{x^3 +\ x^2\quad\quad\quad} \\
-4x^2 + 2x \\
\underline{-4x^2 - 4x} \\
6x + 7 \\
\underline{6x + 6} \\
1
\end{array}
$$

The remainder is 1, so:

$$\frac{x^3 - 3x^2 + 2x + 7}{x + 1} = x^2 - 4x + 6 + \frac{1}{x + 1}$$

Practice questions F

1 Carry out algebraic division on:

(a) $6x^3 + 5x^2 - 17x - 6$ divided by $x + 2$

(b) $2x^3 - x^2 - 5x - 2$ divided by $x + 1$

(c) $3x^3 - 4x^2 - 5x + 2$ divided by $x + 1$

(d) $x^3 - 7x + 6$ divided by $x - 1$

[Leave a space for the missing x^2-term.]

You can check your answers by multiplying back or by using the method of identities.

Quadratic functions

Apart from a linear relationship, the quadratic is the simplest way of connecting two variables x and y, generally written in the form

$$y = ax^2 + bx\ + c$$

We shall be looking at various properties of this particular way of linking two variables, starting with an alternative form for the connecting equation.

Completing the square

OCR P1 5.1.2 (a)

Suppose we take quite a simple quadratic function, say:

$$f(x)\ = x^2 + 4x + 5$$

and try and find a linear function whose square is $f(x)$. We can see that this is not quite possible – the square of $x + 2$ is close,

$$(x + 2)^2 = x^2 + 4x + 4$$

but we need to add 1 to give $f(x)$. Similarly, if we took another function,

$$g(x) = x^2 - 6x + 6$$

the square of $x - 3$ is close,

$$(x - 3)^2\ = x^2 - 6x + 9$$

but now we would have to subtract 3 from this to give $g(x)$. Let's look at these two examples:

$$f(x)\ = x^2 + 4x + 5\ = (x + 2)^2 + 1$$
$$g(x)\ = x^2 - 6x + 6\ = (x - 3)^2 - 3$$

You can see that the constant in the bracket is chosen to be half of the x-coefficient in the standard form, so that the coefficients of x^2 and x are correct. An adjustment then has to be made to the square of this constant so that we have the constant in the standard form. For example:

$$h(x) = x^2 - 8x + 25$$

Halve the x-coefficient, which gives -4, so the bracket is $(x-4)^2$. This gives

$x^2 - 8x + 16$, so we have to add 9 to get $h(x)$:

$$h(x) = x^2 - 8x + 25 = (x-4)^2 + 9$$

If the coefficient of x^2 is something other than 1, we can make the working easier if we rearrange the quadratic first of all. For example, if $f(x) = 2x^2 + 8x + 7$, we take the coefficient of the x^2-term, which is 2, out of the first two terms: $f(x) = 2(x^2 + 4x) + 7$. We then complete the square for the expression in the bracket: $f(x) = 2[(x + 2)^2 - 4] + 7$ and finally remove the square bracket:

$$f(x) = 2(x+2)^2 - 8 + 7 = 2(x+2)^2 - 1$$

Similarly if the coefficient of x^2 is negative:

$$\begin{aligned}
g(x) &= 4 - 6x - 3x^2 \\
&= 4 - 3[x^2 + 2x] \qquad \text{(remember to change the sign} \\
&= 4 - 3[(x + 1)^2 - 1)] \qquad \text{of the } x\text{-term)} \\
&= 4 - 3(x + 1)^2 + 3 \\
&= 7 - 3(x + 1)^2
\end{aligned}$$

Practice questions G

1   Complete the square for the following:

   (a) $x^2 - 2x + 3$ (b) $x^2 + 6x + 7$ (c) $x^2 - 6x + 2$ (d) $x^2 + 3x + 4$

   (e) $2x^2 + 8x + 11$ (f) $3x^2 - 12x + 7$ (g) $3 - 2x - x^2$ (h) $11 + 4x - x^2$

   (i) $5 - 6x - 3x^2$ (j) $11 + 8x - 2x^2$

Graphs of quadratic functions

OCR P1 5.1.3 (f),(g)

You probably already know what the graph of $y = x^2$ looks like, a symmetrical curve with the minimum point at the origin.

Figure 1.1

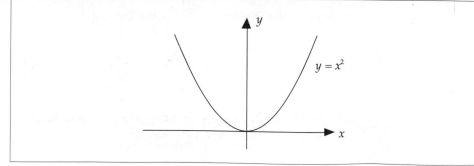

By changing the equation slightly, we can shift the curve around without changing its basic shape. For example, $y = x^2 + 1$ has a graph the same as $y = x^2$ but shifted 1 unit in the y-direction (see Fig. 1.2).

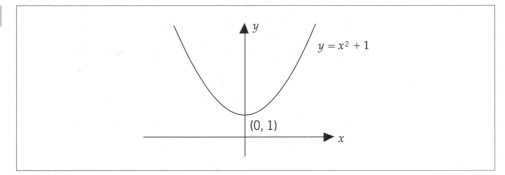

The minimum point is now at (0, 1). Similarly, subtracting something from the x^2 shifts the curve downwards. $y = x^2 - 2$, for example, has the graph shifted –2 units in the y-direction, and its minimum point is now at (0, –2).

Figure 1.3

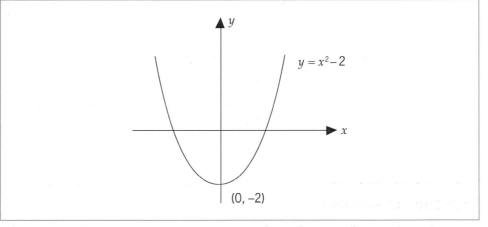

There is a different type of transformation where the term that is squared changes, for example $y = (x - 2)^2$. This shifts the curve in the x-direction but in the opposite sense to what one might expect, so that in fact the graph of $y = (x - 2)^2$ has shifted the graph of $y = x^2$ in the positive direction by 2 units, and the minimum point is now at (2, 0).

Figure 1.4

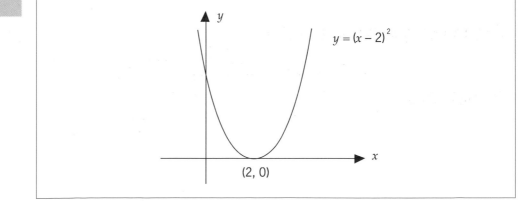

In a similar fashion, the graph of $y = (x + 1)^2$ is a shift of –1 in the x-direction, minimum point at (–1, 0) (see Fig. 1.5).

Figure 1.5

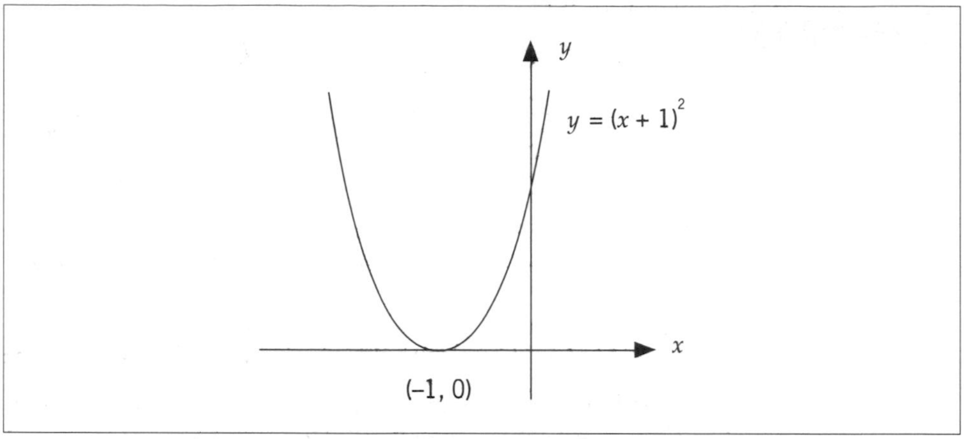

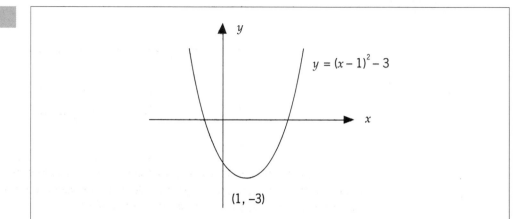

You can also have combinations of these transformations: the graph of
$y = (x - 1)^2 - 3$ for example has a shift of 1 in the x-direction and a shift of -3 in
the y–direction, with the minimum point at $(1, -3)$.

Figure 1.6

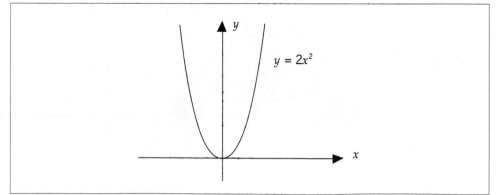

If we multiply the x^2 by a constant, we change the gradient: if the constant is
greater that 1, the curve is steeper. For example, $y = 2x^2$ looks something like the
one shown in Fig. 1.7.

Figure 1.7

If the constant is less than 1, however, the curve is shallower. The graph of
$y = \frac{x^2}{3}$ looks something like the one shown in Fig. 1.8.

Figure 1.8

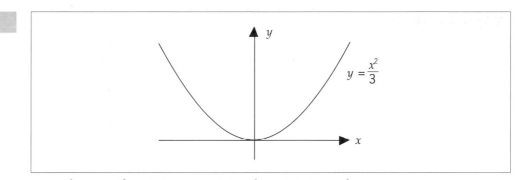

In both cases, the minimum point at the origin is unchanged.

Finally to conclude this section on transformations of the graph $y = x^2$, we have the graph of $y = -x^2$, which is simply the same but upside down, i.e. a reflection in the x-axis. What was previously the minimum point becomes the maximum point.

Figure 1.9

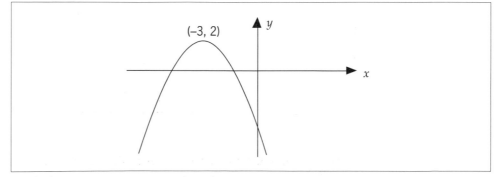

This can also be combined with the other transformations! With a series of transformations the trick is to start as close as possible to the original x^2 term and work outwards. So for example, the graph of

$$y = 2 - (x + 3)^2$$

starts

x^2	$\rightarrow$	$(x + 3)^2$	shift -3 x-direction
$(x + 3)^2$	$\rightarrow$	$-(x + 3)^2$	reflection in x-axis
$-(x + 3)^2$	$\rightarrow$	$2 - (x + 3)^2$	shift $+2$ y-direction

giving the final graph

Figure 1.10

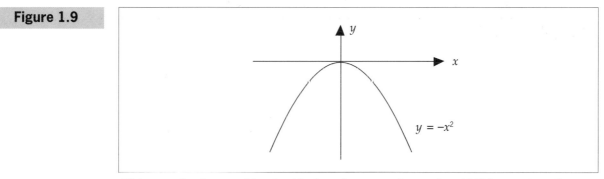

with the maximum point at $(-3, 2)$.

Practice questions H

1   Sketch the graphs corresponding to the following equations, giving the co rdin tes of the minimum or maximum point, as appropriate:

(a) $y = x^2 - 1$ (b) $y = x^2 + 2$

(c) $y = x^2 - c$ [c constant]

(d) $y = (x - 1)^2$ (e) $y = (x + 3)^2$

(f) $y = (x - c^2)$ [c constant]

(g) $y = 1 - x^2$ (h) $y = -2 - x^2$

(i) $y = -(x + 2)^2$ (j) $y = -(x - 1)^2$

(k) $y = 3 + (x - 1)^2$ (l) $y = 2 - (x + 2)^2$

(m) $y = (x + 1)^2 - 5$ (n) $y = 1 - (x - 3)^2$

(o) $y = 3x^2$ (p) $y = \dfrac{x^2}{2}$

(q) $y = 1 + 2x^2$ (r) $y = 3 - \dfrac{x^2}{3}$

Linear inequalities

OCR P1 5.1.2 (c)

The rules for inequalities are the same as those for equations except for one little twist. You can add or subtract anything, provided you do the same to both sides, and you can also multiply or divide both sides by a constant factor, provided this factor is *positive*. But if the factor is negative, the inequality signs have to be *reversed*.

Example

Solve the inequalities:

(a) $3x - 2 > 5$ (b) $5 + 7x < 1$ (c) $2x - 4 < 5x + 2$

(d) $3 - 2x < 9$ (e) $\dfrac{x}{2} + 3 > 1$ (f) $-4 < 2 - x < 4$

Solution

(a) $3x - 2 \;>\; 5$ add 2 to both sides

       $3x \;>\; 7$ divide both sides by 3

       $x \;>\; \dfrac{7}{3}$

(b) $5 + 7x \;<\; 1$ subtract 5

       $7x \;<\; -4$ $\div$ by 7

       $x \;<\; -\dfrac{4}{7}$

(c) $2x - 4 \;<\; 5x + 2$ $-2x$

       $-4 \;<\; 3x + 2$ -2

       $-6 \;<\; 3x$ $\div$ by 3

       $-2 \;<\; x$ i.e. $x > -2$

(d) $3 - 2x \;<\; 9$ -3

       $-2x \;<\; 6$ $\div$ by -2

Here we want to divide by -2, so the inequality sign is now reversed, i.e. $<$ becomes $>$

       $x \;>\; -3$

(e) $\dfrac{x}{2} + 3 \;>\; 1$ -3

       $\dfrac{x}{2} \;>\; -2$ $\times$ by 2

       $x \;>\; -4$

(f)
$$-4 < 2 - x < 4 \qquad -2$$
$$-6 < -x < 2 \qquad \times \text{ by } -1 \text{ (reverses)}$$
$$6 > x > -2$$
i.e. $\quad -2 < x < 6$

Practice questions I

1 Find the values of x for which:

(a) $4x + 3 < 8$ (b) $3x - 2 < 5$ (c) $3x + 4 < 5x + 7$ (d) $3(x - 2) < 6$

(e) $\dfrac{x + 1}{2} < 5$ (f) $4 - 3x < -1$ (g) $-1 < x + 3 < 4$ (h) $-7 < 5 - 2x < 7$

Quadratic inequalities

OCR P1 5.1.2 (c)

If we square a negative number such as -4, we end up with a positive number, in this case $(-4)^2 = 16$. Keeping this in mind, we can see that there are two possible regions where $x^2 > 9$: one is the obvious one, $x > 3$ and the other comes from the above property to give the alternative solution of $x < -3$.

In fact we can generalise this and say that the inequality $x^2 > a^2$ is true either when $x > a$ or $x < -a$. If the inequality sign is reversed, $x^2 < a^2$, we have to reverse the inequality signs in both the solutions to give the pair $x < a$ and $x > -a$. These can be written together as $-a < x < a$.

| **Example** | Solve the inequalities: |

       (a) $\quad x^2 < 4$ (b) $\quad x^2 > 5$ (c) $\quad 2x^2 + 1 < 7$

| **Solution** | (a)   This is simply $\quad -2 < x < 2$ |

(b) $\sqrt{5}$ means the positive square-root of 5, so the solution here is written

$$x > \sqrt{5} \quad \text{or} \quad x < -\sqrt{5}$$

(c) $2x^2 + 1 < 7 \qquad\qquad -1$

$\quad\;\; 2x^2 < 6 \qquad\qquad\quad\; \div \text{ by } 2$

$\quad\;\; x^2 < 3$

$\quad\;\; \Rightarrow \quad -\sqrt{3} < x < 3$

We can apply the same result for more complicated expressions: $(x - 3)^2 < 16$ for example has the solution

$$-4 < x - 3 < 4 \qquad\qquad +3$$

$$-1 < x < 7$$

This could be part of a question which involves completing the square.

Example	Express $f(x) = x^2 - 6x + 5$ in the form $(x + a)^2 + b$, stating the values of the constants a and b.

Hence:

(a) solve the equation $f(x) = 0$

(b) solve the inequality $f(x) > 9$.

Solution	$f(x) = x^2 - 6x + 5 \quad = (x - 3)^2 - 9 + 5$

$$= (x - 3)^2 - 4$$

so $a = -3$ and $b = -4$.

(a) $f(x) = 0 \Rightarrow \quad (x - 3)^2 - 4 \quad = 0$

$$(x - 3)^2 \qquad = 4$$

$$x - 3 \qquad = \pm 2$$

$\Rightarrow$ either $x = 3 + 2 = 5$ or $x = 3 - 2 = 1$.

(b) $f(x) > 9 \Rightarrow (x - 3)^2 - 4 > 9$

$$(x - 3)^2 > 13$$

$\Rightarrow$ either $x - 3 > \sqrt{13}$ i.e. $x > 3 + \sqrt{13}$

or $x - 3 < -\sqrt{13}$ \qquad i.e. $x < 3 - \sqrt{13}$

Practice questions J

1 Solve the inequalities:

(a) $x^2 > 4$

(b) $x^2 < 16$

(c) $x^2 - 3 > 0$

(d) $x^2 + 1 > 0$ (Trick!)

(e) $3 - x^2 < 1$

(f) $(x + 2)^2 > 9$

(g) $(x - 1)^2 < 6$

(h) $2x^2 - 7 > 3$

(i) $(2x - 1)^2 < 25$

(j) $(3x + 1)^2 + 7 < 13$

Graphical method

Sometimes the quadratic is given in factorised form: we can then make a quick sketch and read off the solution from the graph.

Example	Solve the inequalities:

(a) $(x - 2)(x + 5) < 0$ \qquad (b) $(4 - x)(x - 3) < 0$

Solution	(a) Our sketch is:

Figure 1.11

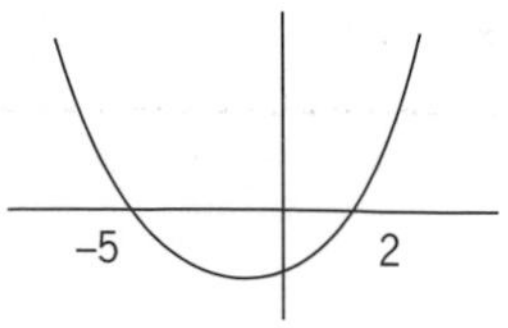

and we want $f(x) < 0$, i.e. we want the values of x for which the curve is below the x-axis. The solution is $-5 < x < 2$.

Figure 1.12

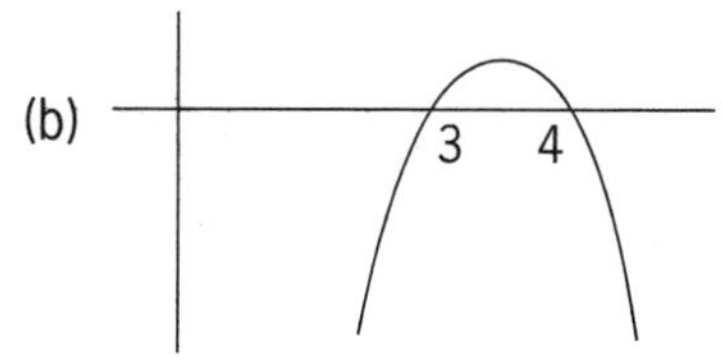

Again, we want the values of x for which the curve is below the x-axis:
$x < 3$ or $x > 4$.

We can use this method even when the quadratic does not factorise: in this case, though, we have first to use the quadratic formula to find the coordinates of the points where the curve crosses the x-axis.

Example

Use the quadratic formula to solve the equation:

$$f(x) = x^2 + 7x - 1 = 0$$

giving your answers to 2 decimal places. Hence solve the inequality $f(x) > 0$.

Solution

Using the formula, $x = \dfrac{-7 \pm \sqrt{49 + 4}}{2}$

$\Rightarrow \quad x = 0.14 \quad$ or $\quad x = -7.14$ (2 d.p.)

The x^2-coefficient is positive, so a sketch could be:

Figure 1.13

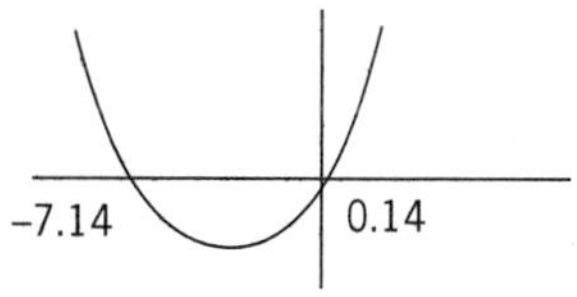

giving the solution of the inequality as $x > 0.14$ or $x < -7.14$. (If you are not familiar with this formula, it is covered immediately after the practice questions.)

Practice questions K

1 Solve the inequalities:

  (a) $(x + 1)(x - 2) < 0$ (b) $(x + 3)(x + 4) > 0$

  (c) $(1 - x)(1 + x) > 0$ (d) $(2 - x)(3 - x) > 0$

2 Factorise and hence solve the inequalities:

  (a) $x^2 - 5x + 6 < 0$ (b) $x^2 - x - 12 > 0$

  (c) $x^2 + 2x \geq 8$ (d) $x^2 - 3x \geq 4$

  (e) $2 - x - x^2 < 0$ (f) $4x + 21 \geq x^2$

3 Given that $f(x) = x^2 - 3x - 5$, use the quadratic formula to solve the equation $f(x) = 0$ giving your solutions correct to 2 decimal places. Solve also the inequality $f(x) < 0$.

Solving quadratic equations

OCR P1 5.1.2 (c)

We saw in the section on polynomials how we can factorise certain quadratics. If the equation we have to solve is in the form $f(x) = 0$ and $f(x)$ is a quadratic function which factorises, the solution is quite straightforward. Since the factors are multiplied together and the product is zero, one of them must be zero. From this we can find the two values of x which satisfy the equation.

Example	Solve the equation: $x^2 - 2x - 8 = 0$

Solution	Factorise first of all, $(x - 4)(x + 2) = 0$

Then either $\qquad x - 4 = 0 \Rightarrow x = 4$

or $\qquad\qquad x + 2 = 0 \Rightarrow x = -2$

Unfortunately, this method does not always work, because not all quadratics will factorise nicely with whole numbers. If we have already put a quadratic function into the completed square form we can sometimes solve related equations quite quickly. Suppose, for example, we had to answer the following question:

Example	Let $f(x) = x^2 + 6x + 7$

(a)   put $f(x)$ into completed square form

(b)   solve the equation $f(x) = 4$

Solution	(a) $f(x) = x^2 + 6x + 7$

$$= (x + 3)^2 - 9 + 7$$

$$= (x + 3)^2 - 2$$

(b)   If $f(x) = 4$, then $(x + 3)^2 - 2 = 4$

$$\Rightarrow (x + 3)^2 = 6$$

We are going to take the square root of both sides.

Remember the ±! This is one of the most common mistakes made by 'A' level students.

$$(x + 3)^2 = 6 \quad \Rightarrow \quad x + 3 = \pm\sqrt{6}$$

$$\Rightarrow \quad x = -3 \quad \pm\sqrt{6}$$

so $x = -3 + \sqrt{6}$ or $x = -3 - \sqrt{6}$

Unless the question asks for a certain number of decimal places or significant figures, we can leave the answers as they are. Note that for any quadratic equation, we expect *two* solutions.

Practice questions L

1. Solve the following equations by the method of factorisation:

   (a) $x^2 - 3x = 0$ (b) $3x^2 + 8x = 0$

   (c) $9x^2 + 12x = 0$ (d) $x^2 + x - 6 = 0$

   (e) $x^2 + x - 30 = 0$ (f) $x^2 - 9 = 0$

   (g) $2x^2 - x - 1 = 0$ (h) $4x^2 - x - 3 = 0$

   (i) $6x^2 - 7x - 3 = 0$ (j) $5x^2 + 19x - 4 = 0$

2. Solve the following equations by first completing the square:

   (a) $x^2 - 4x + 2 = 0$ (b) $x^2 + 6x + 7 = 0$

   (c) $2x^2 - 8x + 5 = 0$ (d) $4 - 2x - x^2 = 0$

   (e) $11 - 6x - 3x^2 = 0$

The quadratic formula

In practice, the numbers occurring in a quadratic equation can be rather awkward and it would take quite a while to find the solution by first completing the square. To save us trouble, there is a formula which will work for any quadratic equation: we just have to put in the figures and out the other end come the two solutions. Even though we have this last resort, we should still be able to factorise and complete the square!

What follows is the proof of the quadratic formula. It is part of the requirement for the Edexcel P1 specification that you know it, so try and read through it and make sure that before you take the exam you can reproduce it!

Suppose we have the general quadratic equation:

$$ax^2 + bx + c = 0$$

where a, b and c are numbers. If we divide both sides of the equation by a, so that the coefficient of x^2 is 1, we have:

$$x^2 + \frac{b}{a}x + \frac{c}{a} = 0 \quad \ldots\ldots \quad [1]$$

If we want to put the expression on the left into the alternative form, the bracket will have to be $\left(x + \frac{b}{2a}\right)^2$ which means that the constant is $\frac{c}{a} - \left(\frac{b}{2a}\right)^2$

$$x^2 + \frac{b}{a}x + \frac{c}{a} = \left(x + \frac{b}{2a}\right)^2 + \frac{c}{a} - \left(\frac{b}{2a}\right)^2$$

so $\left(x + \frac{b}{2a}\right)^2 + \frac{c}{a} - \left(\frac{b}{2a}\right)^2 = 0$ from [1]

i.e. $\left(x + \frac{b}{2a}\right)^2 = \left(\frac{b}{2a}\right)^2 - \frac{c}{a}$

$$= \frac{b^2}{4a^2} - \frac{c}{a}$$

$$= \frac{b^2 - 4ac}{4a^2}$$

Take square roots of both sides, and then:

$$x + \frac{b}{2a} = \pm\sqrt{\frac{b^2 - 4ac}{4a^2}}$$

$$= \pm \frac{\sqrt{b^2 - 4ac}}{2a}$$

i.e. $$x = \frac{-b}{2a} \pm \frac{\sqrt{b^2 - 4ac}}{2a}$$

$$= \frac{-b \pm \sqrt{b^2 - 4ac}}{2a}$$

This is very important and is emphasised below:

The roots of the quadratic equation

$$ax^2 + bx + c = 0$$

are $$x = \frac{-b \pm \sqrt{b^2 - 4ac}}{2a}$$

You may have met this before – but anyway it means that you can now find the roots for *any* quadratic equation by putting the coefficients into this formula. Let's see how it works, for example with:

$$2x^2 - 5x - 9 = 0$$

Here $a = 2$, $b = -5$ and $c = -9$. (Be *very* careful of signs when using the formula.)

Substituting, $$x = \frac{-(-5) \pm \sqrt{(-5)^2 - 4(2)(-9)}}{2(2)}$$

$$= \frac{5 \pm \sqrt{25 + 72}}{4} = \frac{5 \pm \sqrt{97}}{4}$$

so $$x = \frac{5 + \sqrt{97}}{4} \quad \text{or} \quad x = \frac{5 - \sqrt{97}}{4}$$

i.e. $x = 3{\cdot}71$ or $x = -1{\cdot}21$ (correct to two decimal places)

Practice questions M

1 Solve the following equations giving your answer correct to two decimal places:

(a) $x^2 + 4x + 1 = 0$ (b) $x^2 - 7x + 9 = 0$

(c) $x^2 - 3x - 3 = 0$ (d) $2x^2 + 3x - 1 = 0$

(e) $4x^2 + 8x + 1 = 0$

2 Solve the following equations, giving your answer in exact form (i.e. involving square roots):

(a) $x^2 + 6x + 2 = 0$ (b) $x^2 - 5x + 1 = 0$

(c) $2x^2 + 3x - 7 = 0$ (d) $3x^2 + 8x + 2 = 0$

(e) $4x^2 + 2 = 5$

Discriminant

OCR P1 5.1.2 (b)

There are times when we use the quadratic formula and the solution as it works out doesn't make sense. For example, to solve the equation

$$x^2 + 2x + 3 = 0$$

we put $$x = \frac{-2 \pm \sqrt{2^2 - 4(1)(3)}}{2} = \frac{-2 \pm \sqrt{-8}}{2}$$

and here we come to a stop: there is no real number which on being squared gives –8 or in fact any negative number. So to have *real* solutions to a quadratic equation we need the expression inside the square-root not to be negative, i.e.

$$\Delta = b^2 - 4ac \geq 0$$

This quantity is called the *discriminant* and can be written Δ (capital Greek D) for short.

There are three separate cases for the discriminant:

(a) $\Delta > 0$.

For example, using the formula for the equation

$$x^2 - x - 6 = 0$$

$$x = \frac{1 \pm \sqrt{(-1)^2 - 4(1)(-6)}}{2} = \frac{1 \pm \sqrt{25}}{2} = \frac{1 \pm 5}{2}$$

Here, the positive value of the discriminant means that the equation has *two* real roots (which are 3 and –2).

Geometrically, it means that the graph of the quadratic function $y = x^2 - x - 6$ crosses the x-axis (where $y = 0$) in two distinct places (at $x = -2$, $x = 3$).

Figure 1.14

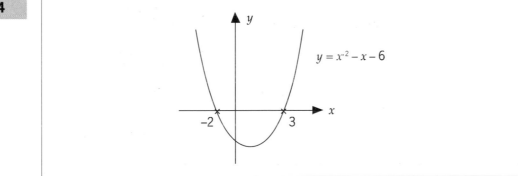

(b) $\Delta = 0$

For example, solving the equation

$$x^2 - 6x + 9 = 0$$

$$x = \frac{6 \pm \sqrt{36 - 36}}{2} = \frac{6 \pm 0}{2}$$

Since adding or subtracting zero makes no difference, there is only *one* value of x which satisfies the equation, $x = 3$.

Geometrically, it implies that the curve just *touches* the x-axis at the single root (see Fig. 1.15).

Figure 1.15

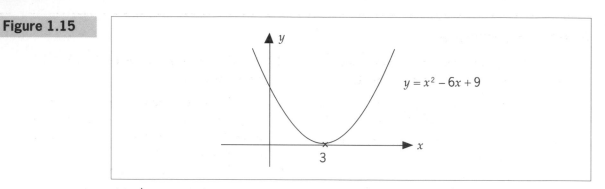

(c) $\Delta < 0$. This we have already seen with the example $x^2 + 2x + 3 = 0$. There are *no* real roots. It means that the graph of $y = x^2 + 2x + 3$ doesn't touch or cross the x-axis at all.

Figure 1.16

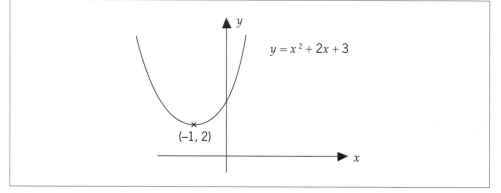

In summary:

Table 1.1

$\Delta = b^2 - 4ac$	Number of real roots
> 0	2
= 0	1
< 0	0

We can use this to find the value, or range of values, of a constant which allows a certain number of solutions to a quadratic equation.

Example

Given the equation:

$$x^2 + ax + 16 = 0$$

find

(a) the values of a for which the equation has a single root

(b) the range of values of a for which the equation has no real root.

Solution

(a) We want $\Delta = 0$, i.e.

$$(a)^2 - 4\,(1)\,(16) = 0$$

$$\Rightarrow a^2 = 64 \quad \text{and} \quad a = \pm 8$$

(b) Here $\Delta < 0$, i.e.

$$a^2 - 64 < 0 \quad \Rightarrow a^2 < 64$$

$$\Rightarrow -8 < a < 8$$

Practice questions N

1 Find the range of values of a for which the following equations have two real roots:

(a) $ax^2 + 2x + 3 = 0$ (b) $x^2 + ax + 4 = 0$

(c) $x^2 + 8x + a = 0$ (d) $x^2 - 2ax + 3a = 0$

2 Find the values of k for which the following equations have equal roots

(a) $kx^2 + 4x + 8 = 0$ (b) $x^2 - 3x + k = 0$

(c) $x^2 - 2kx + 3k = 0$ (d) $(kx + 1)^2 = 8x$

3 Find the range of values for p for which the following equations have no real roots

(a) $2px^2 + 5x + 3 = 0$

(b) $2x^2 + (3 - p) x + p + 3 = 0$

(c) $x^2 + (3p - 4) x + (2p + 8) = 0$

(d) $(2p + 1) x^2 - 3px + 2p = 4$

4 Find the values of k for which $9x^2 + kx + 4$ is a perfect square.

5 Find the range of values of p for which the graph of $y = px^2 + 8x + p - 6$ crosses the x-axis. State also the values of p for which the x-axis is a tangent to the curve.

6 Find the range of values of p so that $px^2 + 8x + 10 - p$ is always positive.

7 Show that $x^2 - 2px + q > 0$ when $q > p^2$. If $q = p^2 - 1$ solve the equation $x^2 - 2px + q = 0$, giving your answer in terms of p.

8 Find the conditions necessary for p and q so that $y = p (x + 2)^2 + q$ does not meet the x-axis. When does it touch?

Simultaneous equations

OCR P1 5.1.2 (d)

When the two equations are linear, we select the simpler of the two equations and make one of the variables the subject. This can then be substituted into the other equation, solved and the corresponding value of the other variable calculated.

Example

Find values of x and y such that:

$$3x - 4y = -1$$

and $\quad x + 2y = 13$

Solution

We rearrange the bottom equation to give $x = 13 - 2y$ and substitute this into the top equation:

$$3(13 - 2y) - 4y = -1$$
$$39 - 6y - 4y = -1$$
$$39 - 10y = -1$$
$$40 = 10y$$
$$y = 4$$

Putting this into $x = 13 - 2y$ gives $x = 13 - 2 \times 4 = 5$.

So the solution is $x = 5$, $y = 4$.

The same method is used when one of the equations is linear and the other quadratic; in this case we invariably rearrange the linear equation.

Example

Find values for s and t which satisfy the simultaneous equations:

$$5s + t = 17$$
$$5s^2 + t^2 = 49$$

Solution

We rearrange the top equation to give:

$$t = 17 - 5s$$

and substitute this into the lower equation. Now it's just a question of some accurate algebra, and care with the signs:

$$5s^2 + (17 - 5s)^2 = 49$$
$$5s^2 + 289 - 170s + 25s^2 = 49$$

Take everything to one side:

$$5s^2 + 289 - 170s + 25s^2 - 49 = 0$$

combining and rearranging:

$$30s^2 - 170s + 240 = 0$$

We can usually divide through by something or other, here by 10:

$$3s^2 - 17s + 24 = 0$$
$$(3s - 8)(s - 3) = 0$$

i.e. $s = \dfrac{8}{3}$ or $s = 3$

Putting this back into the rearranged top equation gives:

$$t = 17 - 5 \times \frac{8}{3} = \frac{11}{3} \quad \text{or} \quad t = 17 - 5 \times 3 = 2$$

We have the two solutions:

$$\left(s = \frac{8}{3}, t = \frac{11}{3}\right) \quad \text{or} \quad (s = 3, t = 2)$$

If we look at the graphs of the two equations, we can see that in fact what we are doing is finding the cooordinates of the points where the two graphs cross.

Example

Find the coordinates of the points of intersection of the line $y = x + 1$ and the curve $y = x^2 - 1$.

Solution

Here the rearranging of the linear equation has already been done for us: all we have to do is equate the two expressions for y, i.e.

$$x + 1 = x^2 - 1$$
$$\Rightarrow \quad x^2 - x - 2 = 0$$
$$(x - 2)(x + 1) = 0$$
$$\Rightarrow \quad \text{either} \quad x = 2 \text{ and so } y = 2 + 1 = 3$$
$$\text{or} \quad x = -1 \text{ and so } y = -1 + 1 = 0$$

the two points of intersection are (2, 3) and (−1, 0)

Practice questions O

1 Solve the following simultaneous equations

(a) $2y - x = 1$
$xy + x^2 = 26$

(b) $3x^2 + y^2 = 28$
$3x + y = 8$

(c) $2x + y = 8$
$4x^2 + 3y^2 = 52$

(d) $2x - y = 5$
$x^2 - xy + y^2 = 7$

(e) $x + 2y = 1$
$x^2 + 4y^2 = 13$

(f) $2x - y = 2$
$x^2 + y^2 = 8$

(g) $2x + y = 7$
$xy = 6$

(h) $2x + 3y = 10$
$\dfrac{2}{x} + \dfrac{3}{y} = 5$

2 Find the points where the given line meets the given curve.

(a) $y = 2x - 1$ meets $y = x^2 - 5x + 9$

(b) $2x + 3y = 14$ meets $xy = 4$

(c) $y = 2x + 1$ meets $y^2 - xy = 3$

(d) $y = 2x + 3$ meets $xy + 20 = 5y$

(e) $y = x - 4$ meets $y^2 + 17 = 2x^2$

(f) $3x + 2y = 4$ meets $x^2 + 2y^2 = 6$

(g) $x + y = 3$ meets $x^2 + 2y^2 = 6$

(h) $2x - 5y + 17 = 0$ meets $xy = 6$

3 The line $3x + 4y = 15$ intersects the curve $2xy = 9$ at the points A and B. Find the coordinates of A and B and find the distance AB.

SUMMARY EXERCISE

1 Express the equation $(x - 5)(x - 3) = 6$ in the form $(x - m)^2 = n$ where m and n are positive integers.

Hence write down the exact values of the two irrational roots of this equation.   [AEB 1998]

2 (a) Find, as surds, the roots of the equation
$$2(x + 1)(x - 4) - (x - 2)^2 = 0$$

(b) Hence find the set of values of x for which
$$2(x + 1)(x - 4) - (x - 2)^2 > 0.$$

3 (a) Use algebra to solve $(x - 1)(x + 2) = 18$

(b) Hence, or otherwise, find the set of values of x for which $(x - 1)(x + 2) > 18$.

4 The quadratic equation
$$x^2 + 6x + 1 = k(x^2 + 1)$$
has equal roots. Find the possible values of the constant k.   [AEB 1994]

5 (a) Show that $(x - 2)$ is a factor of
$$f(x) \equiv x^3 + x^2 - 5x - 2$$

(b) Hence, or otherwise, find the exact solutions of the equation $f(x) = 0$.

6 The quadratic equation $x^2 + kx + 36 = 0$ has two different real roots. Find the set of possible values of k.

7 Given that for all values of x,
$$3x^2 + 12x + 5 \equiv p(x + q)^2 + r,$$

(a) find the values of p, q and r.

(b) Hence, or otherwise, find the minimum value of $3x^2 + 12x + 5$.

(c) Solve the equation $3x^2 + 12x + 5 = 0$, giving your answers to one decimal place.

8 Factorise $49x^2 - 21x + 2$.

Hence, or otherwise, solve the equation
$$49y - 21\sqrt{y} + 2 = 0.$$

Give your answers as fractions.

9 (For this question, use the fact that the area of a triangle is $\frac{1}{2}ab\sin C$.)

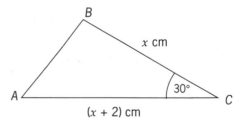

The diagram shows a triangle ABC in which angle $C = 30°$, $BC = x$ cm and $AC = (x + 2)$ cm. Given that the area of triangle ABC is 12 cm^2, calculate the value of x.

10   The diagram shows the graph of $y = x^2 - 2px + p$, where p is a positive constant. The point A is the lowest point on the graph and is given to lie above the x-axis.

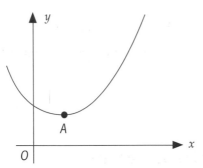

(a)   By completing the square, express the coordinates of A in terms of p. Hence find the set of possible values of p.

(b)   Given that A lies on the straight line with equation $y = 2x - 1$, find the exact value of p.

11   (a)   Find the values of a and b if $(x + 1)$ and $(x + 2)$ are both factors of $x^3 + ax^2 + bx + 4$

(b)   Find the values of the constant A, B and C if $x^3 - 9x^2 + Ax - 24 \equiv (x - 2)(x - B)(x - C)$

(c)   Factorise completely the expression $4x^3 - 13x - 6$ and hence solve the equation $2\left(2x^2 - \dfrac{3}{x}\right) = 13$

12   An isosceles triangle has base 4 cm and height 3 cm. A rectangle $ABCD$ is drawn with its vertices on the sides of the triangle, as shown in the diagram, where $AD = x$ cm and $AB = y$ cm.

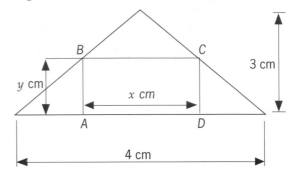

Not to scale

(a)   By considering similar triangles, or otherwise, show that $3x + 4y = 12$.

(b)   Write down an expression for the area of the rectangle. S cm^2, in terms of x alone.

(c)   By completing the square, show that S can be expressed in the form
$$a - \frac{3}{4}(x - b)^2$$
stating the values of the constants a and b.

(d)   State the value of x for which S has its greatest value and write down the maximum value of the area of the rectangle.   [AQA 1996]

SUMMARY   When you have finished this section, you should:

- know what a polynomial is
- be able to add, subtract and multiply polynomials
- be able to factorise quadratics
- know that if a polynomial f(x) is such that f(a) = 0, then (x – a) is a factor of f(x)
- be able to use this to find factors of a polynomial f(x)
- be able to deduce the other factor, either by comparing coefficients or algebraic division
- know how to complete the square for a quadratic
- know how to sketch quadratic functions based on the transformation of the equation of the basic curve
- know how to sketch a quadratic when it is given in the form of factors
- know how to find the minimum or maximum points of these curves
- know how to solve inequalities, both linear and quadratic
- know how to solve quadratic equations when the function can be factorised

- be familiar with the formula for solving quadratic equations and be able to use it
- know how to derive this formula
- know how to find unknown coefficients in identities, either by comparing coefficients or substituting suitable values
- know how the discriminant of a quadratic relates to the roots of the associated equation
- know how to solve simultaneous equations and how this relates to the points of intersection of a line and a curve.

ANSWERS

Practice questions A

1 (a) $6x^2 + 12x$ (b) $6x - 3$
   (c) $-8x + 12$ (d) $-2x - 2x^2$
   (e) $3x - 9x^2$ (f) $-20x + 12x^2$

2 (a) $8x^2 - 10x - 3$ (b) $6x^2 - 7x + 2$
   (c) $4x^2 - 9$ (d) $1 + x - 2x^2$
   (e) $2 + 5x - 3x^2$ (f) $12x^2 - 5x - 2$

3 (a) $3x - 6x^2 + 3x^3$ (b) $1 - x - 4x^3$
   (c) $1 - x^3$ (d) $1 - 4x^2 + 4x^3 - x^4$
   (e) $3 - x - 2x^2 - 5x^3 - 3x^4$
   (f) $1 + 2x + 3x^2 + 2x^3 + x^4$

Practice questions B

1 (a) $A = 3, B = 2, C = 4$
   (b) $A = 2, B = 4$
   (c) $A = -1, B = 1, C = 5, D = -1$
   (d) $A = 2, B = 1, C = 3$
   (e) $A = 2, B = -5, C = 10$
   (f) $A = -2, B = 60, C = 3$

Practice questions C

1 (a) $3x(1 - 2x)$ (b) $4y(x + 2)$
   (c) $pq(q^2 - p)$ (d) $4x^2(3 + 4x)$
   (e) $3(2p - 3q)$ (f) $x(1 + 5x)$

2 (a) $(x + 1)(x + 2)$ (b) $(x + 2)(x + 4)$
   (c) $(x - 3)(x - 4)$ (d) $(x - 2)(x - 3)$
   (e) $(x - 4)(x + 2)$ (f) $(x + 5)(x - 2)$
   (g) $(x - 7)(x + 3)$ (h) $(x - 5)^2$
   (i) $(x + 3)^2$ (j) $(x + 5)(x - 4)$

3 (a) $(2x - 1)(x + 2)$ (b) $(3x + 1)(2x + 1)$
   (c) $(2x + 1)(4x - 3)$ (d) $(3x - 2)(x + 3)$
   (e) $(6x - 1)(x + 2)$ (f) $(2x + 1)(3x - 4)$

Practice questions D

1 (a) $f(2) = 0, f(-2) = 0, f(3) = 0$
   (b) $f(1) = 0, f(2) = 0, f(-3) = 0$
   (c) $f(1) = 0, f(-2) = 0, f(4) = 0$

2 (a) $(x + 1)(x - 2)^2$ (b) $(x - 2)(x - 3)(x - 4)$

3 (a) 5 (b) -28 (c) 0

Practice questions E

1 (a) $a = 3, b = -2, c = 10$
   (b) $a = 1, b = -1, c = 6$
   (c) $a = 3, b = -1, c = 5$
   (d) $a = 1, b = -2, c = -5$

2 (a) $2x^2 + x + 5$ (b) $3x^2 - 3x + 7$
   (c) $3x^2 + 4x - 2$ (d) $4x^2 + 2x + 3$

3 $f(1) = 0, x^2 - 3x - 3$

4 $f(+3) = 0, 2x^2 + 3x + 1 \Rightarrow f(x) = (x - 3)(2x + 1)(x + 1)$

5 (a) $(x - 1)(x + 2)(x - 3)$
   (b) $(x + 1)(x - 2)(3x - 1)$
   (c) $(x - 2)^2(x + 4)$
   (d) $(x + 1)(2x - 1)(x + 2)$
   (e) $(x + 1)(2x - 3)(x + 4)$
   (f) $(x + 1)(2x - 3)(x - 3)$
   (g) $x(x - 3)(x + 2)$
   (h) $(x + 2)^2(x - 3)$
   (i) $(x + 1)(2x - 1)(x - 5)$
   (j) $(x + 1)(x - 2)^2$

Practice questions F

1 (a) $6x^2 - 7x - 3$ (b) $2x^2 - 3x - 2$
   (c) $3x^2 - 7x + 2$ (d) $x^2 + x - 6$

Practice questions G

1   (a) $(x-1)^2 + 2$

    (b) $(x+3)^2 - 2$

    (c) $(x-3)^2 - 7$

    (d) $(x+\frac{3}{2})^2 + \frac{7}{4}$

    (e) $2(x+2)^2 + 3$

    (f) $3(x-2)^2 - 5$

    (g) $4 - (x+1)^2$

    (h) $15 - (x-2)^2$

    (i) $8 - 3(x+1)^2$

    (j) $19 - 2(x-2)^2$

Practice questions H

1   (a)

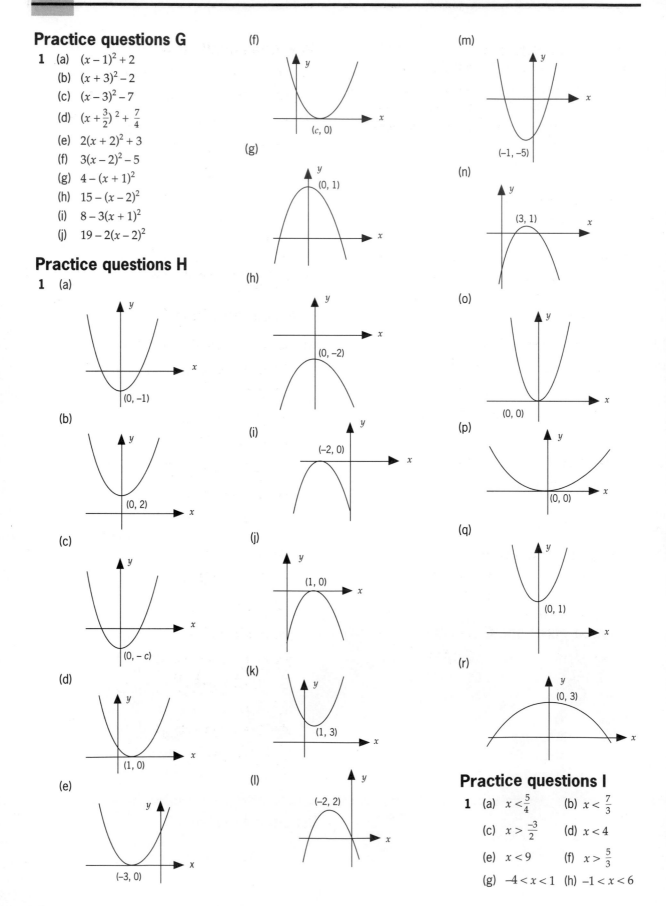

(b)

(c)

(d)

(e)

(f)

(g)

(h)

(i)

(j)

(k)

(l)

(m)

(n)

(o)

(p)

(q)

(r)

Practice questions I

1   (a) $x < \frac{5}{4}$ (b) $x < \frac{7}{3}$

    (c) $x > \frac{-3}{2}$ (d) $x < 4$

    (e) $x < 9$ (f) $x > \frac{5}{3}$

    (g) $-4 < x < 1$ (h) $-1 < x < 6$

Practice questions J

1 (a) $x > 2$ or $x < -2$ (b) $-4 < x < 4$

    (c) $x > \sqrt{3}$ or $x < -\sqrt{3}$ (d) All values

    (e) $x > \sqrt{2}, x < -\sqrt{2}$ (f) $x > 1$ or $x < -5$

    (g) $1 - \sqrt{6} < x < 1 + \sqrt{6}$ (h) $x > \sqrt{5}, x < -\sqrt{5}$

    (i) $-2 < x < 3$

    (j) $-\frac{1}{3}(\sqrt{6} + 1) < x < \frac{1}{3}(\sqrt{6} - 1)$

Practice questions K

1 (a) $-1 < x < 2$ (b) $x < -4$ or $x > -3$

    (c) $-1 < x < 1$ (d) $x > 3$ or $x < 2$

2 (a) $2 < x < 3$ (b) $x > 4$ or $x < -3$

    (c) $x \le -4$ or $x \ge 2$ (d) $x \ge 4$ or $x \le -1$

    (e) $x < -2$ or $x > 1$ (f) $-3 \le x \le 7$

3 $-1.19, 4.19 : -1.19 < x < 4.19$

Practice questions L

1 (a) $0, 3$ (b) $0, \frac{-8}{3}$ (c) $0, \frac{-4}{3}$

    (d) $-3, 2$ (e) $-6, 5$ (f) $3, -3$

    (g) $\frac{-1}{2}, 1$ (h) $1, \frac{-3}{4}$ (i) $\frac{-1}{3}, \frac{3}{2}$

    (j) $\frac{1}{5}, -4$

2 (a) $2 \pm \sqrt{2}$ (b) $-3 \pm \sqrt{2}$

    (c) $2 \pm \sqrt{\frac{3}{2}}$ (d) $-1 \pm \sqrt{5}$

    (e) $-1 \pm \sqrt{\frac{14}{3}}$

Practice questions M

1 (a) $0.27, 3.73$ (b) $1.70, 5.30$

    (c) $-0.79, 3.79$ (d) $-1.78, 0.28$

    (e) $-1.87, -0.13$

2 (a) $-3 \pm \sqrt{7}$ (b) $4 \pm 2\sqrt{3}$

    (c) $\pm \frac{1}{2}\sqrt{3}$ (d) $\frac{-4 \pm \sqrt{10}}{3}$

    (e) $\pm \frac{1}{2}\sqrt{3}$

Practice questions N

1 (a) $a < \frac{1}{3}$ (b) $a > 4$ or $a < -4$

    (c) $a < 16$ (d) $a < 0$ or $a > 3$

2 (a) $k = \frac{1}{2}$ (b) $k = \frac{9}{4}$ (c) $0, 3$ (d) 2

3 (a) $p > \frac{25}{24}$ (b) $-1 < p < 15$

    (c) $-\frac{4}{9} < p < 4$ (d) $p < -\frac{4}{7}$ or $p > 4$

4 ± 12

5 $-2 < p < 8$; -2 and 8

6 $2 < p < 8$

7 $p \pm 1$

8 $pq > 0$ or $p = 0$ and $q \ne 0$.
Touches when $q = 0$

Practice questions O

1 (a) $(4, \frac{5}{2})$ and $(-\frac{13}{3}, -\frac{5}{3})$

    (b) $(1, 5)$ and $(3, -1)$

    (c) $(\frac{5}{2}, 3)$ and $(\frac{7}{2}, 1)$

    (d) $(2, -1)$ and $(3, 1)$

    (e) $(3, -1)$ and $(-2, \frac{3}{2})$

    (f) $(2, 2)$ and $(-\frac{2}{5}, -\frac{14}{5})$

    (g) $(\frac{3}{2}, 4)$ and $(2, 3)$

    (h) $(\frac{1}{2}, 3)$ and $(4, \frac{2}{3})$

2 (a) $(2, 3)$ and $(5, 9)$

    (b) $(6, \frac{2}{3})$ and $(1, 4)$

    (c) $(\frac{1}{2}, 2)$ and $(-2, -3)$

    (d) $(1, 5)$ and $(\frac{5}{2}, 8)$

    (e) $(-11, -15)$ and $(3, -1)$

    (f) $(2, -1)$ and $(\frac{2}{11}, \frac{19}{11})$

    (g) $(2, 1)$

    (h) $(\frac{3}{2}, 4)$ and $(-10, -\frac{3}{5})$

3 $(2, \frac{9}{4})$ and $(3, \frac{3}{2}) : \frac{5}{4}$

Indices

You are probably already familiar with expressions like x^2 and x^3 and know that the numbers 2 and 3 (the *indices*, plural of *index*) tell you how many times you should multiply x (the base) by itself, i.e. $x^2 = x \times x$ and $x^3 = x \times x \times x$. In general x^m, where m is some positive integer, means x has been multiplied by itself m times.

In fact, this idea can be extended to include all kinds of numbers as indices, not just positive integers. We can, for example, have x^{-2}, $x^{\frac{1}{2}}$, x^0, $x^{-1.78}$ and so on, and all these have their own particular meaning.

There are only a few rules for indices and they appear quite simple. This is deceptive – for many students this is a tricky topic and causes quite a bit of trouble. At the same time it is extremely important for later work. So it really is worth spending a while on this section, making sure that you completely understand the rules and are confident that you can answer the corresponding examples.

Adding and multiplying indices

OCR P1 5.1.1 (a)

One distinction that you have to make clear is between $x^m \times x^n$ and $(x^m)^n$. To take a numerical example, what is the difference between $2^2 \times 2^3$ and $(2^2)^3$? Well, 2^2 means 2×2 and 2^3 means $2 \times 2 \times 2$, so $2^2 \times 2^3$ means $(2 \times 2) \times (2 \times 2 \times 2)$, which is five 2's multiplied together, $2^5 = 32$. On the other hand $(2^2)^3$ means 3 2^2's multiplied together, $2^2 \times 2^2 \times 2^2$. Since each 2^2 is two 2's multiplied together, we have 6 2's in all, so $(2^2)^3 = 2^6 = 64$. In general, $x^m \times x^n$ means a total of $(m + n)$ x's multiplied together, and $(x^m)^n$ means $(m \times n)$ x's multiplied together.

① $\qquad x^m \times x^n = x^{m+n}$

② $\qquad (x^m)^n = x^{mn}$

To take another example, $x^5 \times x^3 = x^8$, whereas $(x^5)^3 = x^{15}.$

Notes

1. Remember that although we don't write it in full usually, x actually means x^1 so that $x \times x^5 = x^1 \times x^5 = x^6$.

2. Be careful when multiplying something like 2×3^5: a common mistake is to say that it is the same as 6^5 – it isn't! (We would need to multiply $2^5 \times 3^5$ to get 6^5.)

Negative indices

If we try and apply the definition that x^m means multiplying x by itself m times to the expression 2^{-3}, we don't seem to have anything meaningful: how can 2 be multiplied by itself a negative number of times? An alternative approach is to use the rule that $x^m \times x^n = x^{m+n}$ and see what happens if n is negative.

For example, taking m to be 5, n to be –3 and using 2 as the base, we have $2^5 \times 2^{-3} = 2^2$. Then what value must we assign to 2^{-3} if the equation is to be true? We have $2^5 = 32$ and $2^2 = 4$,

so $32 \times 2^{-3} = 4 \implies 2^{-3} = \dfrac{4}{32} = \dfrac{1}{8} = \dfrac{1}{2^3}$

So it looks as though things work out if we take 2^{-3} to be the same as $\dfrac{1}{2^3}$. In fact, whatever the base x and the index n, things work if we take x^{-n} to mean the same as $\dfrac{1}{x^n}$.

$$\text{③} \qquad x^{-n} = \dfrac{1}{x^n}$$

Since multiplying by $\dfrac{1}{x^n}$ is the same as *dividing* by x^n (e.g. multiplying by a half is the same as dividing by 2), we have

$$x^m \div x^n = x^m \times \dfrac{1}{x^n} = x^m \times x^{-n} = x^{m-n}$$

and we have the following rule

$$\text{④} \qquad x^m \div x^n = x^{m-n}$$

In particular, $x^{-1} = \dfrac{1}{x^1} = \dfrac{1}{x}$ is called the *reciprocal* of x: $3^{-1} = \dfrac{1}{3}$ and $p^{-1} = \dfrac{1}{p}$, for example. If x is a fraction, this is turned upside down:

$$\left(\dfrac{2}{3}\right)^{-1} = \dfrac{3}{2} \qquad \left(\dfrac{x}{y}\right)^{-1} = \dfrac{y}{x} \qquad \text{and} \quad \left(\dfrac{1}{2}\right)^{-1} = \dfrac{2}{1} = 2$$

Multiplying expressions

When multiplying or dividing expressions involving constants, e.g. $3x^2 \times 4x^{-3}$, we deal with the constants first and then deal with the variables

$$3x^2 \times 4x^{-3} = 12x^2 \times x^{-3}$$
$$= 12x^{-1}$$

Here are a few examples using these rules ...

| **Example** | Express in the form x^k the following expressions: |

(a) $x^4 \times x^6$ (b) $(x^4)^6$ (c) $(x^{-3})^2$ (d) $x^{-4} \times x^5$

(e) $\dfrac{1}{x^3}$ (f) $x^7 \times \dfrac{1}{x^3}$ (g) $x^8 \div x^4$ (h) $x^2 \div x^{-4}$

Solution	(a) $x^4 \times x^6 = x^{10}$	rule ①	(e) $\dfrac{1}{x^3} = x^{-3}$	rule ③
	(b) $(x^4)^6 = x^{24}$	rule ②	(f) $x^7 \times \dfrac{1}{x^3} = x^7 \times x^{-3} = x^4$	rule ①
	(c) $(x^{-3})^2 = x^{-6}$	rule ②	(g) $x^8 \div x^4 = x^{8-4} = x^4$	rule ④
	(d) $x^{-4} \times x^5 = x^1 = x$	rule ①	(h) $x^2 \div x^{-4} = x^{2-(-4)} = x^6$	rule ④

| **Example** | Simplify the following: |
| | (a) $4x^4 \times 2x^2$ (b) $5x^2 \times 2x^{-4}$ (c) $8x \div 4x^3$ (d) $(2x)^3 \times x$ |

Solution	(a) $4x^4 \times 2x^2 = 8x^4 \times x^2 = 8x^6$
	(b) $5x^2 \times 2x^{-4} = 10x^2 \times x^{-4} = 10x^{-2}$
	(c) $8x \div 4x^3 = 2x \div x^3 = 2x^1 \div x^3 = 2x^{1-3} = 2x^{-2}$
	(d) $(2x)^3 \times x = 8x^3 \times x = 8x^4$

Index of zero

What are the consequences if we allow one of the indices in the rule $x^m \times x^n = x^{m+n}$ to be zero? Take x to be 2, m to be 5 and n zero. Our rule states that $2^5 \times 2^0 = 2^{5+0} = 2^5$ i.e. $32 \times 2^0 = 32$, so that for this to be true, 2^0 must have the value of one. This has also to be true in general, so that except where $x = 0$,

$$⑤ \quad x^0 = 1$$

Fractional indices

We can further extend the values that the indices m and n can take if we allow them to be fractions and see what this leads to. We can take rule ②, $(x^m)^n = x^{mn}$ with $m = \dfrac{1}{2}$ and $n = 2$, which gives

$$(x^{\frac{1}{2}})^2 = x^{\frac{1}{2} \times 2} = x^1 = x$$

so when we square $x^{\frac{1}{2}}$, we end up with x. We already have a term for this, the *square-root* of x, written $\sqrt{x}$,

so $x^{\frac{1}{2}} = \sqrt{x}$

Similarly, taking $m = \dfrac{1}{3}$ and $n = 3$, we have

$$(x^{\frac{1}{3}})^3 = x^{\frac{1}{3} \times 3} = x^1 = x$$

and so $x^{\frac{1}{3}}$ must be the cube-root of x, $\sqrt[3]{x}$.

In general:

$$\textcircled{6} \quad x^{\frac{1}{p}} = \sqrt[p]{x}$$

When the index is more complicated, something like $\frac{3}{4}$, we can use rule ② to separate the top and the bottom of the fraction

$$x^{\frac{3}{4}} = x^{3 \times \frac{1}{4}} = (x^3)^{\frac{1}{4}}, \qquad \text{or alternatively}$$

$$x^{\frac{3}{4}} = x^{\frac{1}{4} \times 3} = (x^{\frac{1}{4}})^3$$

If we needed to evaluate $16^{\frac{3}{4}}$ for example, we would probably use the second alternative:

$$16^{\frac{3}{4}} = (16^{\frac{1}{4}})^3 = 2^3 = 8, \qquad \text{instead of} \quad 16^{\frac{3}{4}} = (16^3)^{\frac{1}{4}} = \dots$$

since you might know that the fourth-root of 16, $\sqrt[4]{16}$. or $16^{\frac{1}{4}}$, is 2, but 16^3 is a big number which you then have to square-root!

Example

Evaluate the following, without using a calculator:

(a) $27^{\frac{1}{3}}$ (b) 3^{-3} (c) $\left(\frac{4}{5}\right)^{-1}$ (d) $\left(\frac{2}{3}\right)^{-2}$

(e) $8^{\frac{2}{3}}$ (f) $64^{-\frac{1}{3}}$ (g) $25^{\frac{3}{2}}$ (h) $16^{-\frac{3}{4}}$

Solution

(a) $27^{\frac{1}{3}} = \sqrt[3]{27} = 3$ (b) $3^{-3} = \frac{1}{3^3} = \frac{1}{27}$

(c) $\left(\frac{4}{5}\right)^{-1} = \frac{5}{4}$ (d) $\left(\frac{2}{3}\right)^{-2} = \left(\frac{3}{2}\right)^2 = \frac{9}{4}$

(e) $8^{\frac{2}{3}} = \left(8^{\frac{1}{3}}\right)^2 = 2^2 = 4$ (f) $64^{-\frac{1}{3}} = \frac{1}{64^{\frac{1}{3}}} = \frac{1}{\sqrt[3]{64}} = \frac{1}{4}$

(g) $25^{\frac{3}{2}} = (25^{\frac{1}{2}})^3 = 5^3 = 125$ (h) $16^{-\frac{3}{4}} = \frac{1}{16^{\frac{3}{4}}} = \frac{1}{(16^{\frac{1}{4}})^3} = \frac{1}{2^3} = \frac{1}{8}$

Example

Express the following in the form x^k:

(a) $x \times x^{\frac{1}{2}}$ (b) $x^{\frac{2}{3}} \times x^{-\frac{1}{3}}$ (c) $\left(x^{-\frac{1}{2}}\right)^2$

(d) $\sqrt[3]{x^4}$ (e) $(\sqrt{x})^5$ (f) $\frac{1}{\sqrt{x}}$

Solution

(a) $x \times x^{\frac{1}{2}} = x^1 \times x^{\frac{1}{2}} = x^{1 + \frac{1}{2}} = x^{\frac{3}{2}}$ (b) $x^{\frac{2}{3}} \times x^{-\frac{1}{3}} = x^{\frac{2}{3} - \frac{1}{3}} = x^{\frac{1}{3}}$

(c) $\left(x^{-\frac{1}{2}}\right)^2 = x^{-1}$ (d) $\sqrt[3]{x^4} = (x^4)^{\frac{1}{3}} = x^{\frac{4}{3}}$

(e) $(\sqrt{x})^5 = (x^{\frac{1}{2}})^5 = x^{\frac{5}{2}}$ (f) $\frac{1}{\sqrt{x}} = \frac{1}{x^{\frac{1}{2}}} = x^{-\frac{1}{2}}$

Have a go at some of these.

Practice questions A

1 Evaluate the following without using a calculator:

(a) 4^{-1} (b) 3^{-2} (c) $\left(\frac{3}{4}\right)^{-1}$ (d) $\left(\frac{3}{2}\right)^{-3}$ (e) $\left(\frac{1}{2}\right)^{-2}$

(f) $\sqrt{2^4}$ (g) $8^{\frac{1}{3}}$ (h) $8^{-\frac{2}{3}}$ (i) $(2^3)^{-1}$ (j) 5^0

(k) $16^{\frac{1}{4}}$ (l) $(2^{-2})^2$ (m) $\frac{1}{9^{-1}}$ (n) $(-8)^{\frac{1}{3}}$ (o) $\left(\frac{1}{49}\right)^{\frac{1}{2}}$

(p) $\left(\frac{27}{8}\right)^{\frac{2}{3}}$ (q) $\frac{2^{-1}}{3}$ (r) $\frac{3}{4^{-1}}$ (s) $\sqrt[3]{8^2}$ (t) $\sqrt{4^3}$

2 Express the following in the form $y = x^k$:

(a) $x^2 \times x^4$ (b) $(x^4)^3$ (c) $(x^{-2})^2$ (d) $(x^4)^{\frac{1}{2}}$ (e) $\frac{1}{\sqrt{x}}$

(f) $\frac{1}{x^{-1}}$ (g) $\sqrt{x^4}$ (h) $x^{\frac{1}{2}} \times x^{\frac{1}{2}}$ (i) $x^{\frac{1}{2}} \times x$ (j) $(x^{\frac{1}{2}})^3$

(k) $x^5 \div x^2$ (l) $x \div x^{-2}$ (m) $\frac{x^2}{\sqrt{x}}$ (n) $x^2 \sqrt{x}$ (o) $\frac{x \times \sqrt{x}}{x^{-1}}$

(p) $\frac{x}{\sqrt[3]{x}}$

3 Simplify the following:

(a) $2x^2 \times 5x^3$ (b) $x \times 3x^2 \times 4x^3$ (c) $(3x)^2 \div x$ (d) $(2x)^{-2} \times 16x^3$

(e) $8x^3 \div 2x^{-1}$ (f) $(2x^2)^{-1} \times (4x)^{-2}$ (g) $4x^2 \div 8x^5 \times 3x^7$ (h) $(2x)^2 \times (4x)^{-\frac{1}{2}} \times 8x^{\frac{3}{2}}$

Use in algebra

You will find later on when you come to the sections on more advanced calculus that you need to know how to deal with expressions involving indices and fractions. Here are some typical examples.

Example	Expand:
	(a) $(\sqrt{x} + 1)^2$ (b) $x^{\frac{1}{2}} (x^2 - 4)$ (c) $(3\sqrt{x} - 1)(x + 1)$ (d) $(3x + 4) x^{-2}$

Solution	(a) $(\sqrt{x} + 1)^2 = (\sqrt{x})^2 + 2\sqrt{x} + 1 = x + 2\sqrt{x} + 1$
	(b) $x^{\frac{1}{2}} (x^2 - 4) = x^{\frac{5}{2}} - 4x^{\frac{1}{2}}$
	(c) $(3\sqrt{x} - 1)(x + 1) = 3x^{\frac{3}{2}} + 3x^{\frac{1}{2}} - x - 1$
	(d) $(3x + 4) x^{-2} = 3x^{-1} + 4x^{-2}$

Example	Express the following in a form not involving fractions (except in the indices):
	(a) $\frac{(x^2 + 1)}{x}$ (b) $\frac{(x - 2)^2}{\sqrt{x}}$

Solution	(a) $\dfrac{(x^2+1)}{x} = (x^2+1)\ x^{-1} = x + x^{-1}$

(b) $\dfrac{(x-2)^2}{\sqrt{x}} = \dfrac{x^2-4x+4}{\sqrt{x}} = (x^2-4x+4)\ x^{-\frac{1}{2}} = x^{\frac{3}{2}} - 4x^{\frac{1}{2}} + 4x^{-\frac{1}{2}}$

Practice questions B

1   Expand the following expressions:

(a) $\sqrt{x}\,(x-1)$ (b) $(\sqrt{x}-1)^2$ (c) $(x-2)\,x^{-1}$ (d) $(x^2+1)\,x^{-\frac{1}{2}}$

(e) $(2-\sqrt{x}\,)^2$ (f) $(x^{\frac{1}{2}}+1)(x^{-\frac{1}{2}}-1)$ (g) $\left(\sqrt{x}+\dfrac{1}{\sqrt{x}}\right)\left(\sqrt{x}-\dfrac{1}{\sqrt{x}}\right)$

(h) $\left(\sqrt{x}-\dfrac{1}{\sqrt{x}}\right)^2$ (i) $x^2\left(\sqrt{x}-\dfrac{2}{\sqrt{x}}\right)$ (j) $\sqrt{x}\left(x+\dfrac{1}{x}\right)$

2   Express the following in a form not involving fractions (apart from those in the indices):

(a) $\dfrac{x+1}{\sqrt{x}}$ (b) $\dfrac{(x-1)^2}{\sqrt{x}}$ (c) $\dfrac{(x+1)^2}{x^3}$ (d) $\dfrac{\sqrt{x}-1}{x^2}$ (e) $\dfrac{(\sqrt{x}-1)^2}{x}$ (f) $\dfrac{x^2-1}{x^{\frac{1}{3}}}$

Solving equations

We saw before that if $x^p = y$ then $x = y^{\frac{1}{p}}$. For example, $2^3 = 8 \Rightarrow 2 = 8^{\frac{1}{3}} = \sqrt[3]{8}$. So if we have something like $x^3 = 27$, we can solve this by using the above property and writing down $x = 27^{\frac{1}{3}} = \sqrt[3]{27} = 3$. Knowing that we take the reciprocal of the index (i.e. $\dfrac{1}{p}$ if the index is p) we can solve equations with more complicated indices:

Example	Solve $4x^{\frac{2}{3}} = 25$

Solution	$x^{\frac{2}{3}} = \dfrac{25}{4}$

$$\Rightarrow\quad x = \left(\dfrac{25}{4}\right)^{\frac{3}{2}} = \left(\left(\dfrac{25}{4}\right)^{\frac{1}{2}}\right)^3 = \left(\dfrac{5}{2}\right)^3 = \dfrac{125}{8}$$

Notes

1   We can take cube-roots of negative numbers. Since $(-2)^3 = -8$, $(-8)^{\frac{1}{3}} = -2$.

2   We cannot take square-roots of negative numbers. x^2 is always positive for any x other than zero.

3   When we take the square-root of a positive number, we get two solutions. If $x^2 = 25$, then $x = +\sqrt{25}$ or $x = -\sqrt{25}$, i.e. $x = \pm 5$. By definition, both $\sqrt{25}$ and the corresponding $25^{\frac{1}{2}}$ mean $+5$.

4 Be careful with constants: $3x^{-1} = \dfrac{3}{x}$ and $\dfrac{1}{2x} = \dfrac{1}{2}x^{-1}$. It is very common to see

$3x^{-1}$ written as $\dfrac{1}{3x}$ and $\dfrac{1}{2x}$ written as $2x^{-1}$, both of which are incorrect.

Some further examples of these equations.

Example	

Solve the following equations:

(a) $5x^{-1} = 3$ (b) $9x^2 = 4$ (c) $x^3 + 1 = 0$ (d) $x^{-2} = 4$

(e) $x^{-\frac{2}{3}} = 16$ (f) $x^{\frac{1}{4}} = 2$

Solution	

(a) $5x^{-1} = 3 \implies \dfrac{5}{x} = 3 \implies x = \dfrac{5}{3}$

(b) $9x^2 = 4 \implies x^2 = \dfrac{4}{9} \implies x = \pm\dfrac{2}{3}$

(c) $x^3 + 1 = 0 \implies x^3 = -1 \implies x = (-1)^{\frac{1}{3}} = -1$

(d) $x^{-2} = 4 \implies \dfrac{1}{x^2} = 4 \implies x^2 = \dfrac{1}{4} \implies x = \pm\dfrac{1}{2}$

(e) $x^{-\frac{2}{3}} = 16 \implies \dfrac{1}{x^{\frac{2}{3}}} = 16 \implies x^{\frac{2}{3}} = \dfrac{1}{16}$

$$\implies x = \left(\tfrac{1}{16}\right)^{\frac{3}{2}} = \left(\left(\tfrac{1}{16}\right)^{\frac{1}{2}}\right)^3 = \left(\tfrac{1}{4}\right)^3 = \dfrac{1}{64}$$

(f) $x^{\frac{1}{4}} = 2 \implies x = 2^4 = 16$

Try some of these for yourself.

Practice questions C

1 Solve the following equations, where possible:

(a) $9x^{-1} = 2$ (b) $x^{\frac{2}{3}} = 4$ (c) $4x^2 = 1$ (d) $x^{-3} = 8$ (e) $x^{\frac{1}{3}} = 2$ (f) $x^3 + 27 = 0$

(g) $4x^{-2} = 3x^{-1}$ (h) $x^{\frac{1}{2}} = 2x^{-\frac{1}{2}}$ (i) $x^{-\frac{2}{3}} = 9$ (j) $8x^3 + 1 = 0$ (k) $8x = 27x^{-\frac{1}{2}}$ (l) $4x^2 + 9 = 0$

Equations with unknown indices

We can see quite quickly that if $2^x = 8$, then x must have the value of 3. To solve a slightly different equation such as $2^x = 7$ requires some further technique which you will be covering later on in the course, but there are some equations of a particular type which we can solve immediately with the help of the material in this section. These equations have numbers which are all exact powers of the same base, usually 2 or 3. Let's have a look at an example of this type.

Example	Express as powers of 2: (a) 4^{2x} (b) 8^{-x}
	Hence solve the equation $4^{2x} = 2 \times 8^{-x}$

Solution	(a) $4^{2x} = (2^2)^{2x} = 2^{4x}$

(b) $8^{-x} = (2^3)^{-x} = 2^{-3x}$

The equation becomes

$$2^{4x} = 2 \times 2^{-3x} = 2^1 \times 2^{-3x} = 2^{1-3x}$$

Now we have a situation where $2^p = 2^q$ and so p must have the same value as q,

i.e. $4x = 1 - 3x \Rightarrow 7x = 1 \Rightarrow x = \frac{1}{7}$

Practice questions D

1 Express as powers of 3:

(a) 9^{2x} (b) 27^{x-1}

Hence solve the equation $\frac{9^{2x}}{3^2} = 27^{x-1}$

2 Solve the following equations:

(a) $5^{2x} = 25^{1-x}$ (b) $2 \times 4^x = 32$

(c) $\frac{4^{x-1}}{2} = 8^x$ (d) $\sqrt{27^x} = 3 \times 9^x$

Surds

OCR P1 5.1.1 (b)

A surd is the square root of a positive real integer like $\sqrt{21}$ or $\sqrt{3}$. By definition, although in taking a square root we have two possible solutions, one positive and one negative, we take the positive square root, so that $\sqrt{4}$ means 2 and not –2. In fact, if the number inside the square root is not a perfect square the surd is an *irrational* number, i.e. it cannot be written as a fraction. This caused considerable consternation amongst the ancient Greek mathematicians when they realised that $\sqrt{2}$ could never be represented exactly in the form $\frac{a}{b}$. In a later section, we shall be seeing how they proved this. Since these surds are *exact* as they stand, it's frequently preferable to keep them as they are and learn how to manipulate them rather than using a decimal approximation like $\sqrt{3} = 1.732$.

The rules for multiplying and dividing are quite simple:

$$\sqrt{x} \times \sqrt{y} = \sqrt{xy}$$

$$\frac{\sqrt{x}}{\sqrt{y}} = \sqrt{\frac{x}{y}}$$

So that, for example $\sqrt{3} \times \sqrt{5} = \sqrt{15}$ and $\sqrt{21} \div \sqrt{3} = \sqrt{\frac{21}{3}} = \sqrt{7}$. We can use the top rule to simplify the numbers inside the surd as far as possible,

e.g. $\sqrt{50} = \sqrt{2 \times 25} = \sqrt{2} \times \sqrt{25} = 5\sqrt{2}$

and $\sqrt{12} = \sqrt{4 \times 3} = \sqrt{4} \times \sqrt{3} = 2\sqrt{3}$

As with algebraic expressions, we can collect like terms together, so that

$3\sqrt{6} + 5\sqrt{6} = 8\sqrt{6}$ and $10\sqrt{5} - 13\sqrt{5} = -3\sqrt{5}$, for example.

Practice questions E

1 Simplify the following:

   (a) $\sqrt{18}$ (b) $\sqrt{98}$

   (c) $\sqrt{80}$ (d) $\sqrt{200}$

   (e) $\sqrt{32}$ (f) $\sqrt{75}$

2 Simplify and collect like terms:

   (a) $\sqrt{18} + \sqrt{8}$ (b) $2\sqrt{12} + 3\sqrt{75}$

   (c) $3\sqrt{5} + \sqrt{80}$ (d) $\sqrt{50} - \sqrt{32}$

   (e) $\sqrt{200} - 2\sqrt{98}$

   (f) $2\sqrt{80} + 3\sqrt{125} - 4\sqrt{20}$

Rationalising the denominator

We prefer to have any surds in the numerator rather than the denominator if they occur in a fraction: the process of eliminating the surds on the bottom of a fraction is called 'rationalising the denominator'. In the simplest case, we have a single surd as denominator, something like:

$$\frac{6}{\sqrt{2}}$$

We than multiply top and bottom of the fraction by $\sqrt{2}$: by doing this, we square the denominator and clear it of surds:

$$\frac{6}{\sqrt{2}} \times \frac{\sqrt{2}}{\sqrt{2}} \quad = \frac{6\sqrt{2}}{2} \quad = 3\sqrt{2}$$

When the denominator consists of a mixture of numbers and surds, like $2 - \sqrt{3}$ or two different surds, like $\sqrt{5} + 2\sqrt{2}$, we use a different method which relies on the fact that $(a + b)(a - b) = a^2 - b^2$. So if we multiply either of the above expressions by the same expression with an opposite sign in the middle, we end up with the difference of two squares, which eliminates the surds:

$$(2 - \sqrt{3})(2 + \sqrt{3}) \quad = 2^2 - (\sqrt{3})^2$$
$$= 4 - 3 = 1$$
$$(\sqrt{5} + 2\sqrt{2})(\sqrt{5} - 2\sqrt{2}) = (\sqrt{5})^2 - (2\sqrt{2})^2$$
$$= 5 - 8 = -3$$

Be careful when you're squaring mixtures of numbers and surds,

e.g. $(3\sqrt{2})^2 = 3\sqrt{2} \times 3\sqrt{2} = 9 \times 2 = 18$

Let's see how this method of multiplying by the opposite expression clears the denominator of surds.

Example

(a) Express $\dfrac{14}{3 - \sqrt{2}}$ in the form $a + b\sqrt{2}$

(b) Express $\dfrac{5}{2\sqrt{2} - \sqrt{7}}$ in the form $c\sqrt{2} + d\sqrt{7}$

| Solution | (a) Multiply top and bottom of the fraction by the opposite of $3 - \sqrt{2}$, |

i.e. $3 + \sqrt{2}$

$$\frac{14}{3 - \sqrt{2}} \times \frac{3 + \sqrt{2}}{3 + \sqrt{2}} = \frac{14(3 + \sqrt{2})}{3^2 - \sqrt{2}^2} = \frac{14(3 + \sqrt{2})}{9 - 2}$$

$$= \frac{14(3 + \sqrt{2})}{7} = 2(3 + \sqrt{2})$$

$$= 6 + 2\sqrt{2}$$

(b) This time, top and bottom by $2\sqrt{2} + \sqrt{7}$

$$\frac{5}{2\sqrt{2} - \sqrt{7}} \times \frac{2\sqrt{2} + \sqrt{7}}{2\sqrt{2} + \sqrt{7}} = \frac{5(2\sqrt{2} + \sqrt{7})}{(2\sqrt{2})^2 - (\sqrt{7})^2} = \frac{5(2\sqrt{2} + \sqrt{7})}{8 - 7}$$

$$= 5(2\sqrt{2} + \sqrt{7})$$

$$= 10\sqrt{2} + 5\sqrt{7}$$

Practice questions F

1 Rationalise the denominator and simplify where possible:

(a) $\dfrac{12}{\sqrt{3}}$ (b) $\dfrac{20}{\sqrt{5}}$ (c) $\dfrac{16}{\sqrt{8}}$

(d) $\dfrac{5}{\sqrt{2}}$ (e) $\dfrac{3\sqrt{2}}{\sqrt{3}}$ (f) $\dfrac{4\sqrt{7}}{\sqrt{2}}$

(g) $\dfrac{9\sqrt{50}}{2\sqrt{3}}$ (h) $\dfrac{7\sqrt{18}}{3\sqrt{35}}$

2 Rationalise the denominator:

(a) $\dfrac{1}{2 - \sqrt{3}}$ (b) $\dfrac{3}{4 - \sqrt{7}}$ (c) $\dfrac{2\sqrt{2}}{\sqrt{2} + 1}$

(d) $\dfrac{12}{3 - \sqrt{3}}$ (e) $\dfrac{6}{\sqrt{5} - \sqrt{2}}$ (f) $\dfrac{2 + \sqrt{2}}{2 - \sqrt{2}}$

(g) $\dfrac{8\sqrt{5}}{3 - \sqrt{5}}$ (h) $\dfrac{5\sqrt{2} - 10\sqrt{7}}{\sqrt{7} - \sqrt{2}}$

SUMMARY EXERCISE

1 Solve the simultaneous equations

$$\frac{125^x}{25^y} = 625 \quad \text{and} \quad 2 \times 4^x = 32^y$$

2 Without a calculator, find the value of

(a) $\dfrac{3}{\sqrt{2} - 1} - \dfrac{6}{\sqrt{2}}$ (b) $\dfrac{1}{3 - \sqrt{5}} + \dfrac{1}{3 + \sqrt{5}}$

3 Solve the equations

(a) $x^{\frac{2}{3}} = 8x$ (b) $2 \times 4^{x+1} = 8^x$

4 When $x = \dfrac{-1}{2}$, find the value of

(a) $\dfrac{1}{3x}$ (b) $\dfrac{1}{3x^{-2}}$

5 Show that the substitution $y = 2^x$ transforms the equation

$$2(2^x) + 2^{-x} = 3$$

into the quadratic equation

$$2y^2 - 3y + 1 = 0.$$

Hence find the values of x which satisfy the equation

$$2(2^x) + 2^{-x} = 3 \qquad \text{[AQA 1999]}$$

6 It is given that $a^p = 5$ and $a^q = 9$. In each of the following cases, determine the numerical value of the given expression.

(a) a^{p+q} (b) $2a^{-p}$ (c) $a^{2p - \frac{1}{2}q}$

7 By using the substitution $y = x^{\frac{1}{3}}$, solve the equation

$$x^{\frac{2}{3}} - 5x^{\frac{1}{3}} + 6 = 0.$$

8 In this question, use the fact that

$$q^2 = p^2 + r^2 - 2pr \cos Q \quad \text{(cosine rule)}$$

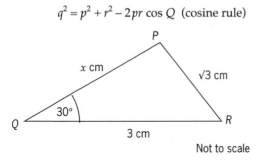

Not to scale

The diagram shows triangle PQR with angle $PQR = 30°$, $QR = 3$ cm, $PR = \sqrt{3}$ cm and $PQ = x$ cm. Given that the exact value of $\cos 30°$ is $\frac{1}{2}\sqrt{3}$, show that the possible values of x are given by the equation

$$x^2 - (3\sqrt{3})x + 6 = 0.$$

Hence find the possible values of x, giving your answers in the form $k\sqrt{3}$, where k is an integer.

9 Solve the equations

(a) $x = \sqrt{x + 9} + 3$

(b) $\sqrt{4y - 9} = 2\sqrt{y} - 1$.

10 Find without a calculator the values of

(a) $\left(5^3\right)^{\frac{2}{3}}$ (b) $4^{-\frac{3}{2}}$ (c) $8^{\frac{4}{3}} \div 8^{\frac{2}{3}}$

11 By means of the substitution $y = 2^x$, find the value of x such that

$$2^{x+2} - 2 = 7 \times 2^{x-1}$$

12 (a) Solve the equation

$$x^2 + (3\sqrt{3})x - 30 = 0$$

giving each of your answers in the form $p\sqrt{3}$, where p is an integer.

(b) Solve the equation

$$3^{\frac{2}{3}} + (3\sqrt{3})z^{\frac{1}{3}} - 30 = 0$$

giving each of your answers in the form $r\sqrt{3}$, where r is an integer

13 (a) (i) Given that $b = a^6$, express $b^{\frac{1}{8}}$ as a power of a.

(ii) Given that

$$c = \frac{\left(a^{\frac{3}{4}} + 5b^{\frac{1}{8}}\right)}{a^{\frac{1}{2}}} \quad \text{and} \quad b = a^6$$

write c in the form ka^p, stating the values of the constants k and p.

(b) Simplify $\left(4a^{\frac{1}{2}}b\right)^2 \div (2a^{-1}b^2)$

14 By substituting $X = 4^x$ and $Y = 3^y$, express the following simultaneous equations in terms of X and Y.

$$4^{x+1} - 3^y = 1$$

$$4^x + 3^y = 1.5$$

Solve the new equations to obtain the values of X and Y. Hence find the values of x and y.

SUMMARY

When you have finished this section, you should

- be clear about the distinction between $x^m \times x^n = x^{m+n}$ and $\left(x^m\right)^n = x^{mn}$

- know that $a^{-b} = \dfrac{1}{a^b}$ and $\dfrac{1}{c^{-d}} = c^d$

- be able to simplify expressions such as $3x^{-2} \times 4x^{\frac{5}{2}}$

- know that $a^0 = 1$ for any a except $a = 0$

- know that $\sqrt{x} = x^{\frac{1}{2}}$ and $\sqrt[3]{x} = x^{\frac{1}{3}}$

- be able to rewrite expressions such as $\sqrt[p]{x^q}$ as $x^{\frac{q}{p}}$

- be able to evaluate expressions like $125^{-\frac{2}{3}}$ without the use of a calculator
- be able to solve certain equations of the form $x^a = b$
- know how to transform equations into a quadratic by means of a suitable substitution
- know how to simplify surds
- know how to rationalise the denominator when this contains surds.

ANSWERS

Practice questions A

1. (a) $\frac{1}{4}$ (b) $\frac{1}{9}$ (c) $\frac{4}{3}$ (d) $\frac{8}{27}$

   (e) 4   (f) 4   (g) 2   (h) $\frac{1}{4}$

   (i) $\frac{1}{8}$ (j) 1   (k) 2   (l) $\frac{1}{16}$

   (m) 9   (n) –2   (o) $\frac{1}{7}$ (p) $\frac{9}{4}$

   (q) $\frac{1}{6}$ (r) 12   (s) 16   (t) 8

2. (a) x^6 (b) x^{12} (c) x^{-4} (d) x^2

   (e) $x^{-\frac{1}{2}}$ (f) x (g) x^2 (h) x

   (i) $x^{\frac{3}{2}}$ (j) $x^{\frac{3}{2}}$ (k) x^3 (l) x^3

   (m) $x^{\frac{3}{2}}$ (n) $x^{\frac{5}{2}}$ (o) $x^{\frac{5}{2}}$ (p) $x^{\frac{2}{3}}$

3. (a) $10x^5$ (b) $12x^6$ (c) $9x$

   (d) $4x$ (e) $4x^4$ (f) $\frac{1}{32}x^{-4}$

   (g) $\frac{3}{2}x^4$ (h) $16x^3$

Practice questions B

1. (a) $x^{\frac{3}{2}} - x^{\frac{1}{2}}$ (b) $x - 2\sqrt{x} + 1$

   (c) $1 - \frac{2}{x}$ (d) $x^{\frac{3}{2}} + x^{-\frac{1}{2}}$

   (e) $4 - 4\sqrt{x} + x$ (f) $x^{-\frac{1}{2}} - x^{\frac{1}{2}}$

   (g) $x - \frac{1}{x}$ (h) $x - 2 + \frac{1}{x}$

   (i) $x^{\frac{5}{2}} - 2x^{\frac{3}{2}}$ (j) $x^{\frac{3}{2}} + x^{-\frac{1}{2}}$

2. (a) $x^{\frac{1}{2}} + x^{-\frac{1}{2}}$ (b) $x^{\frac{3}{2}} - 2x^{\frac{1}{2}} + x^{-\frac{1}{2}}$

   (c) $x^{-1} + 2x^{-2} + x^{-3}$ (d) $x^{-\frac{3}{2}} - x^{-2}$

   (e) $1 - 2x^{-\frac{1}{2}} + x^{-1}$ (f) $x^{\frac{5}{3}} - x^{-\frac{1}{3}}$

Practice questions C

1. (a) $\frac{9}{2}$ (b) 8   (c) $\pm\frac{1}{2}$ (d) $\frac{1}{2}$

   (e) 8   (f) –3   (g) $\frac{4}{3}$ (h) 2

   (i) $\frac{1}{27}$ (j) $-\frac{1}{2}$ (k) $\frac{9}{4}$

   (l) Not possible

Practice questions D

1. (a) 3^{4x} (b) 3^{3x-3}, $x = -1$

2. (a) $\frac{1}{2}$ (b) 2   (c) –3   (d) –2

Practice questions E

1. (a) $3\sqrt{2}$ (b) $7\sqrt{2}$ (c) $4\sqrt{5}$ (d) $10\sqrt{2}$
   (e) $4\sqrt{2}$ (f) $5\sqrt{3}$

2. (a) $5\sqrt{2}$ (b) $19\sqrt{3}$ (c) $7\sqrt{5}$ (d) $\sqrt{2}$
   (e) $-4\sqrt{2}$ (f) $15\sqrt{5}$

Practice questions F

1. (a) $4\sqrt{3}$ (b) $4\sqrt{5}$ (c) $2\sqrt{8} = 4\sqrt{2}$

   (d) $\frac{5}{2}\sqrt{2}$ (e) $\sqrt{6}$ (f) $2\sqrt{14}$

   (g) $\frac{15}{2}\sqrt{6}$ (e) $\frac{1}{5}\sqrt{70}$

2. (a) $2 + \sqrt{3}$ (b) $\frac{4 + \sqrt{7}}{3}$ (c) $2\sqrt{2}(\sqrt{2} - 1)$

   (d) $2(3 + \sqrt{3})$ (e) $2(\sqrt{5} + \sqrt{2})$ (f) $3 + 2\sqrt{2}$

   (g) $2\sqrt{5}(3 + \sqrt{5})$ (e) $-12 - \sqrt{14}$

Trigonometry

When we try use mathematics to model something that occurs in the physical world we find that many phenomena can be described very successfully in terms of trigonometric functions: the progress of waves at sea or the string of a musical instrument vibrating, for example. This section is an introduction to the study of the fundamental trigonometric functions, their graphs and the techniques necessary for calculations which involve them. We begin by looking at an alternative way of measuring angles which is used very widely in more advanced mathematics.

Radians

OCR P2 5.2.1(a)

We usually measure angles in *degrees* – comparing the size of the angle with the 360° that make up a complete revolution.

Figure 3.1

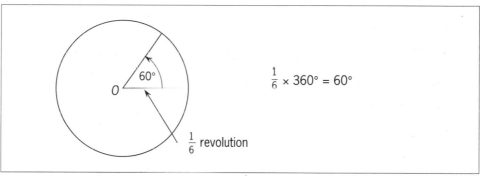

$\frac{1}{6} \times 360° = 60°$

$\frac{1}{6}$ revolution

Another way we can express this is to compare the length of the arc subtending the angle with the perimeter of the whole circle. The units used in this case are called **radians**.

If we take a circle of radius 1, the angle at the centre when the arc length is 0.7 is 0.7 radians; when the arc length is 1.3, the angle is 1.3 radians, etc.

Figure 3.2

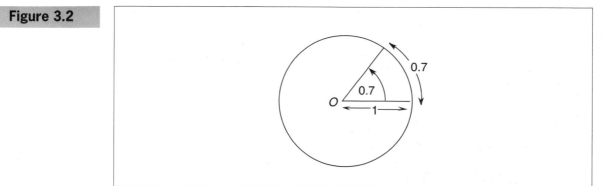

We write one radian as 1 rad or simply as 1 if we know we are dealing with radians. Since the perimeter of the circle is $2\pi r$, it will require 2π radians to complete the revolution, i.e.

$$2\pi \text{ radians} = 360 \text{ degrees}$$

$$\text{and} \quad 1 \text{ radian} = \frac{360°}{2\pi}$$

$$= 57 \cdot 29578...°$$

This is not a very convenient number. With 2π radians equivalent to 360°, π radians = 180°, and so, if the angle in degrees is some multiple of a factor of 180°, we can express it in radians as a multiple of π. Some common angles are:

$$90° = \frac{\pi}{2} \qquad 30° = \frac{\pi}{6} \qquad 120° = \frac{2\pi}{3}$$

Working the other way round:

$$\frac{\pi}{4} = 45° \qquad \frac{\pi}{10} = 18° \qquad \frac{7\pi}{6} = 210°$$

You will need to use radians later on, especially in integration, curve-sketching and solving equations, when there are mixtures of algebraic and trigonometric functions. For the moment, it's important to get used to changing from one system to the other and recognising the common angles.

Practice questions A

1   Express in radians:

   (a) 360°   (b) 60°   (c) 270°

   (d) 540°   (e) 120°   (f) 240°

   (g) 75°   (h) 225°

   (i) 315°   (j) 135°

2   Express in degrees:

   (a) $\dfrac{\pi}{3}$ (b) $\dfrac{3\pi}{4}$ (c) $\dfrac{2\pi}{3}$ (d) $\dfrac{5\pi}{6}$

   (e) $\dfrac{3\pi}{2}$ (f) 4π (g) $\dfrac{5\pi}{4}$ (h) $\dfrac{7\pi}{6}$

   (i) $\dfrac{5\pi}{3}$ (j) $\dfrac{\pi}{10}$ radians.

Arcs and areas of sectors

OCR P2 5.2.1(b)

The perimeter of a circle is $2\pi r$, and its area is πr^2. If we take a sector of a circle, its arc length and area will be a fraction of the whole. For example, half a circle has an arc length of $\frac{1}{2} \times 2\pi r = \pi r$ units and an area of $\frac{1}{2} \times \pi r^2 = \frac{\pi r^2}{2}$ square units.

We measure the fraction by the proportion of the central angle to 360° or 2π radians.

An angle of 60° is then $\frac{60}{360} = \frac{1}{6}$ th of the whole, so a sector of radius 5 cm with a subtended angle of 60° has:

$$\text{arc length} \quad \frac{1}{6} \times 2\pi r = \frac{10\pi}{6} = \frac{5\pi}{3} \text{ cm}$$

$$\text{area} \quad \frac{1}{6} \times \pi r^2 = \frac{25\pi^2}{6} \text{ cm}^2$$

Figure 3.3

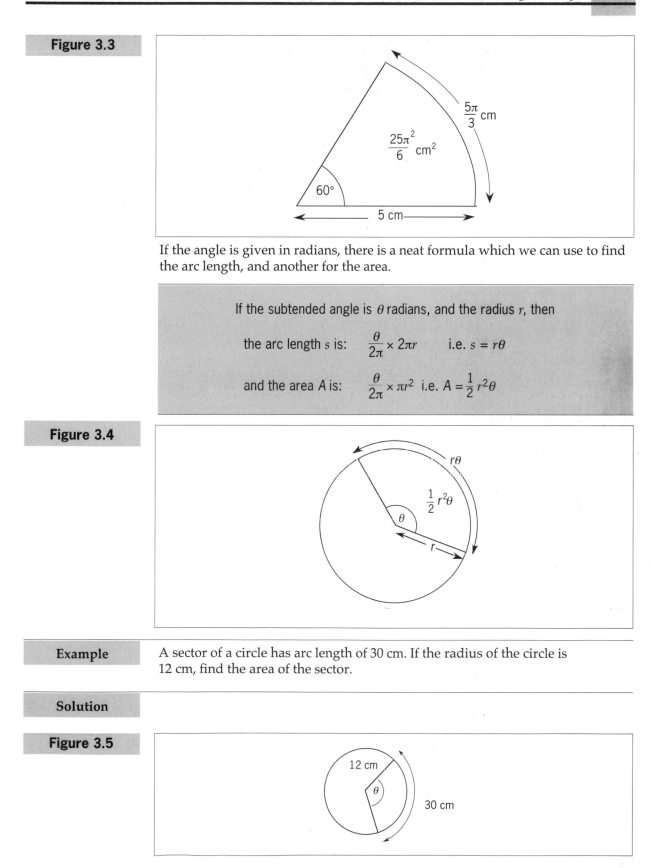

If the angle is given in radians, there is a neat formula which we can use to find the arc length, and another for the area.

If the subtended angle is θ radians, and the radius r, then

the arc length s is: $\dfrac{\theta}{2\pi} \times 2\pi r$ i.e. $s = r\theta$

and the area A is: $\dfrac{\theta}{2\pi} \times \pi r^2$ i.e. $A = \dfrac{1}{2} r^2 \theta$

Figure 3.4

Example

A sector of a circle has arc length of 30 cm. If the radius of the circle is 12 cm, find the area of the sector.

Solution

Figure 3.5

Using the formula for arc length,

$$s = r\theta \Rightarrow 30 = 12\theta \Rightarrow \theta = \frac{30}{12} = \frac{5}{2} \text{ (radians)}$$

Using formula for area,

$$A = \frac{1}{2} r^2\theta = \frac{1}{2} \times 12^2 \times \frac{5}{2} = 180 \text{ cm}^2$$

Practice questions B

1 Find the arc lengths and areas of sectors subtending an angle of:

  (a)   90°   with radius 4 cm

  (b)   210°   with radius 12 cm

  (c) $\frac{3\pi}{4}$ with radius 8 cm

  (d) $\frac{5\pi}{3}$ with radius 9 cm

2 The figure shows a sector of a circle, radius r, and central angle $\frac{\pi}{6}$. If the area is 3π, find r.

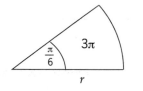

3 AOB is a sector of a circle. The length of the arc AB is 21 and the angle AOB is 3 radians. Find:

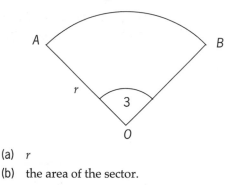

  (a) r

  (b)   the area of the sector.

Arcs and chords

OCR P2 5.2.1(b)

Corresponding to the **arc** (curve) PQ of a sector is the **chord** (straight line) PQ.

Figure 3.6

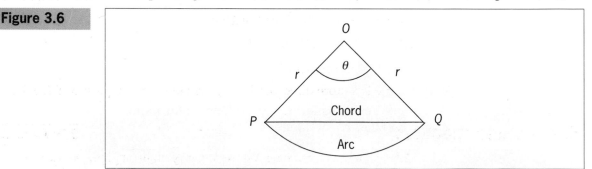

Questions on this part of the syllabus often involve figures with a mixture of arcs/chords and sectors/triangles, so we need to know how to calculate the various lengths and areas. Since the angle θ is measured in radians, your calculator needs to be in *radian* mode when dealing with any of the trigonometric functions sin, cos or tan.

Chord *PQ*

If we look at the triangle OPQ in the figure above, we can see that it is isosceles, since both OP and OQ are radii. This means that dropping a perpendicular from

O to N on PQ will bisect the angle at O and bisect the base PQ, forming two (equal) right-angled triangles.

Figure 3.7

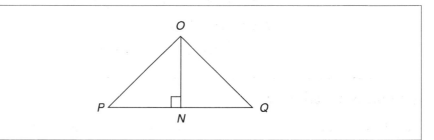

From the triangle PON we can use trigonometry to find PN and hence the length of the chord PQ. For example, suppose $OP = OQ = 10$ cm and $\angle POQ = 1.6$ radians. Then $\angle PON = \frac{1}{2} \times 1.6 = 0.8$ rads.

Figure 3.8

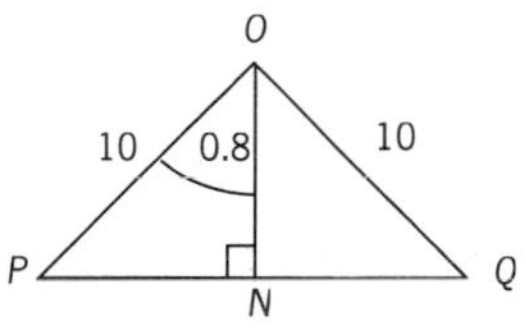

From $\triangle OPN$, $\sin 0.8 = \dfrac{PN}{10}$

$\Rightarrow$ $PN = 10 \sin 0.8 = 7.174$ cm [remember: *radian* mode in your calculator]

So $PQ = 2PN = 14.3$ cm (to 3 s.f.)

From the same triangle, we could find the height ON, using $\cos 0.8 = \dfrac{ON}{10}$

$\Rightarrow$ $ON = 10 \cos 0.8 = 6.97$ cm.

To find the area of $\triangle POQ$, we could then use the formula $\frac{1}{2}$ base $\times$ height for a right-angled triangle, which would give:

Area $\triangle POQ$ $= \frac{1}{2} \times 14.348 \times 6.967$

$= 50.0$ (to 3 s.f.)

In fact, there is a more direct method commonly used in questions like this.

Sectors and triangles

OCR P2 5.2.1(b)

To find the area of any triangle (not necessarily right-angled), we can use the formula

$$\text{Area } \Delta = \frac{1}{2} ab \sin C$$

where a and b are any sides and C is the angle between them. When both sides are radii and the angle between them is θ, this becomes

$\frac{1}{2} \times r \times r \sin \theta = \frac{1}{2} r^2 \sin \theta.$

In particular, for the triangle *POQ*, the are would be

$$\frac{1}{2} \times 10^2 \times \sin 1.6 = 50.0 \text{ (to 3 s.f.)}$$

which is the answer we had before.

Here are two examples using these ideas.

| **Example** | *OAB* is a sector of a circle, radius 12 cm. Find: |

(a)   the perimeter

(b)   the area of the shaded region – this is called a **segment**.

| **Figure 3.9** |

| **Solution** | (a)   Drawing in a perpendicular from *O* onto AB gives |

| **Figure 3.10** |

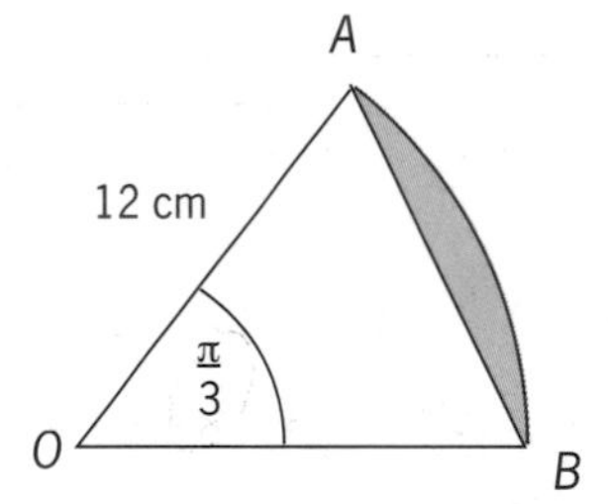

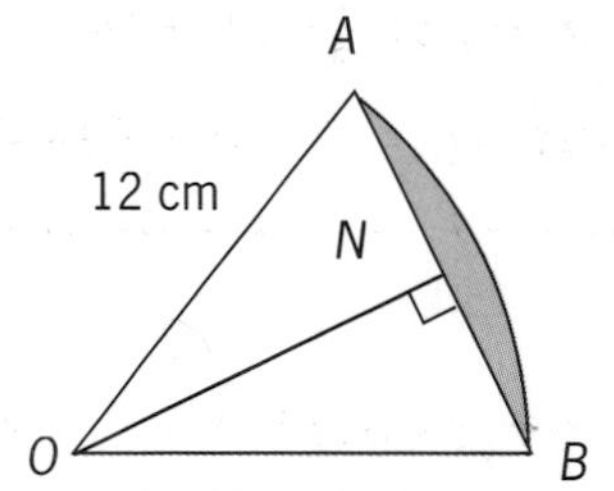

For $\triangle AON$, $\angle AON = \frac{1}{2} \times \frac{\pi}{3} = \frac{\pi}{6}$

$\Rightarrow \quad \sin \angle AON = \dfrac{AN}{12}$

$\Rightarrow \quad AN = 12 \sin \angle AON = 6$ cm

$\Rightarrow \quad$ Chord $AB = 2AN = 12$ cm

Arc $AB = r\theta = 12 \times \dfrac{\pi}{3} = 4\pi$

$\Rightarrow \quad$ Perimeter of shaded region is $12 + 4\pi = 24.6$ cm (to 3 s.f.)

(b)   Area $\triangle AOB = \frac{1}{2} r^2 \sin \theta$

$$= \frac{1}{2} \times 12^2 \times \sin \frac{\pi}{3} = 62.354 \text{ cm}^2$$

Area sector $AOB = \frac{1}{2} r^2 \theta = 75.398$ cm²

Then shaded area is $75.398 - 62.354 = 13.0$ cm² (to 3 s.f.)

Example	OAB is a right-angled triangle with $OA = OB = 8$ cm. Find the area enclosed between the semi-circle with AB as diameter and the arc AB of the circle, centre O and radius OA.

Figure 3.11

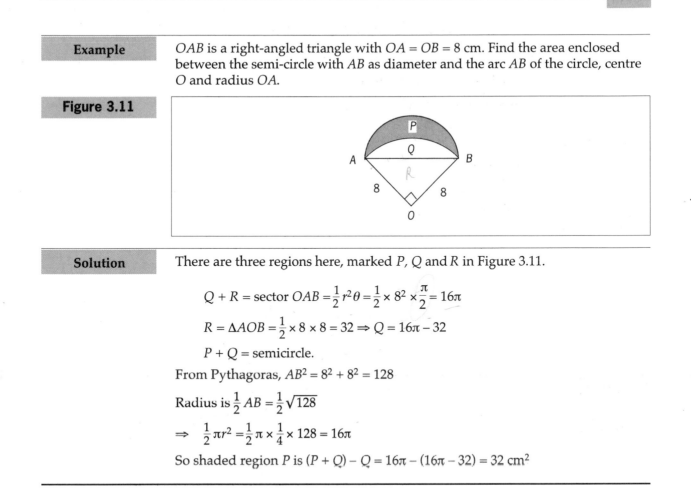

Solution	There are three regions here, marked P, Q and R in Figure 3.11.

$$Q + R = \text{sector } OAB = \frac{1}{2}r^2\theta = \frac{1}{2} \times 8^2 \times \frac{\pi}{2} = 16\pi$$

$$R = \Delta AOB = \frac{1}{2} \times 8 \times 8 = 32 \Rightarrow Q = 16\pi - 32$$

$P + Q = $ semicircle.

From Pythagoras, $AB^2 = 8^2 + 8^2 = 128$

Radius is $\frac{1}{2} AB = \frac{1}{2}\sqrt{128}$

$$\Rightarrow \quad \frac{1}{2}\pi r^2 = \frac{1}{2}\pi \times \frac{1}{4} \times 128 = 16\pi$$

So shaded region P is $(P + Q) - Q = 16\pi - (16\pi - 32) = 32 \text{ cm}^2$

Practice questions C

1 Find the perimeter of the following shaded regions. In each case, curved portions are part of a circle, centre O.

2 For each of the figures in question 1, find the area of the shaded region.

3 Find the area of the following shaded regions:

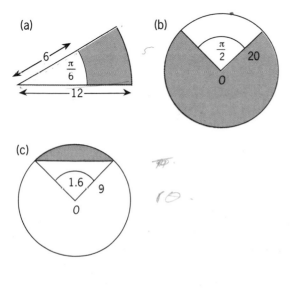

(a)

(b)

(c)

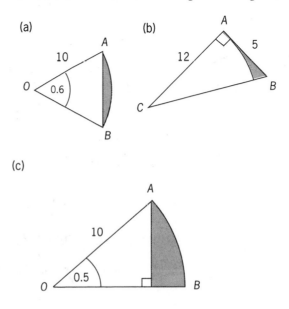

(a)

(b)

(c)

(d)

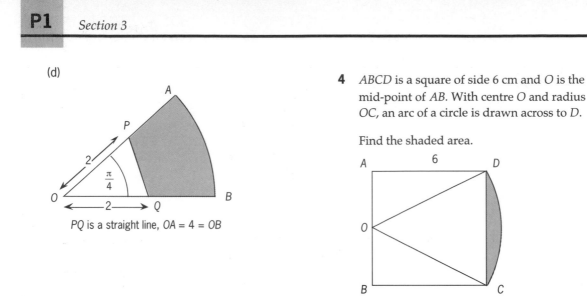

PQ is a straight line, OA = 4 = OB

4 *ABCD* is a square of side 6 cm and *O* is the mid-point of *AB*. With centre *O* and radius *OC*, an arc of a circle is drawn across to *D*.

Find the shaded area.

The sine of any angle

OCR P1 5.1.4 (a)

We're used to the idea that the sine of an angle is the ratio of the opposite side to the hypotenuse in a right-angled triangle. This definition can be extended to give a value for sine of any angle:

> As a point *P* moves round a unit circle with centre the origin 0,
> the value of sin θ, where θ is the angle OP makes with the positive *x*-axis,
> is the *y*-coordinate of the point *P*.

The advantage of this definition is that it removes the restriction on the size of the angle θ. For example, when $\theta = 210°$, the *y*-coordinate of *P* is $-\frac{1}{2}$, and so $\sin 210° = -\frac{1}{2}$.

Figure 3.12

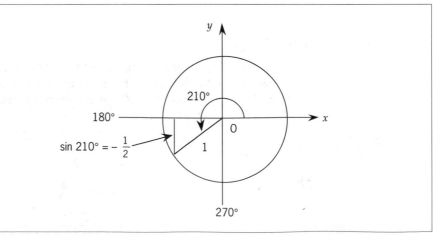

Continuing in this way, we arrive back at the starting point. If we keep going, we have an angle more than 360°, but we ignore the first revolution so that

$$\sin 390° = \sin (390° - 360°) = \sin 30° = \frac{1}{2}$$

Figure 3.13

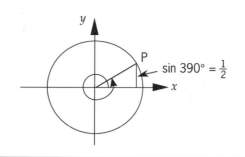

Then, by plotting the angle against the value of the corresponding y-coordinate, we end up with a graph which may be familiar to you.

Figure 3.14

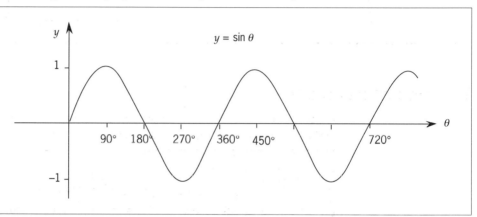

You can see that it passes through the origin, because when θ is zero, P is on the x-axis, i.e. $y = 0$.

It also has a maximum value of 1, and a minimum value of -1 when θ is 90° and 270° respectively, which is when P is right at the top and bottom of the circle. You can see that at 360° the whole graph starts again, repeating itself exactly. This is an example of a *periodic function*, that is a function with values that repeat at regular intervals: the *period* in this case is 360° (or 2π in radians).

The cosine of any angle

OCR P1 5.1.4 (a)

Cos θ has a similar definition, except that this time as the point moves round the circle we take the value of the x-coordinate. This means that after P has reached the top, when θ is 90°, cos θ becomes negative for a while since the x-coordinate of P is negative.

Figure 3.15

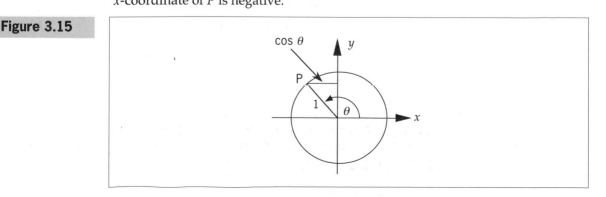

The graph of cos θ is actually exactly the same as sin θ except that it has been shifted back 90°.

Figure 3.16

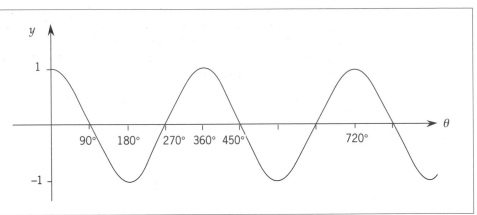

The maximum value is now when θ is 0°, and you can see that the graph is negative when θ is more than 90°, reaching a minimum at 180° until P is at the bottom of the circle at 270° where the x-coordinate is zero. It then increases until a new maximum of 1 when θ is 360° and so on, over and over again.

In fact, both the graph of sin θ and that of cos θ can be continued backwards if we allow the point to travel clockwise, which means the angle is decreasing. The graphs then look like this ...

Figure 3.17

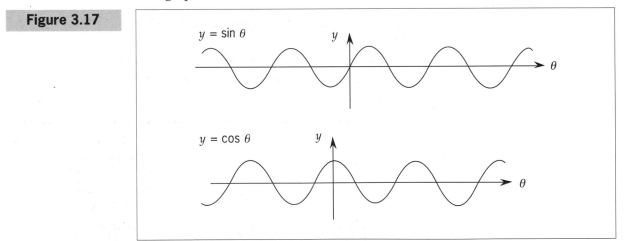

The tangent of any angle

OCR P1 5.1.4 (a)

The final function we have to look at is that of tan θ. It is defined simply as:

$$\tan \theta = \frac{\sin \theta}{\cos \theta}$$

It doesn't look anything like either of the other two graphs, because of the term cos θ as denominator. When cos θ is very small, which means that θ is close to 90°, tan θ becomes very large. Try and find the value of tan 89.99° on your calculator and you'll see this for yourself.

When θ is actually 90°, the denominator of the fraction is zero and the value of tan θ is undefined (your calculator will just display E). To show this on the graph, we put in a dotted line (called an asymptote) at $\theta = 90°$. Just after 90°, cos θ is still very small, but it's now *negative*. Since sin θ is still positive, tan θ will also be negative but extremely large. The graph jumps from the top of the page to the bottom – when the graph has a break like this, we say that the function is *discontinuous*. After 180°, both sin θ and cos θ are negative, so that tan θ becomes positive once more, building up to another asymptote when cos θ is zero again, which is at 270°.

Again, we can continue the graph backwards and we end up with something like this …

Figure 3.18

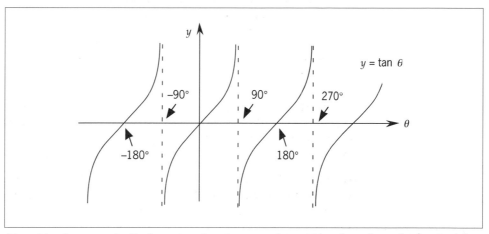

Tan θ is also a periodic function, but its period is 180° rather than the 360° of the other two (or π in radians).

Here's a table to summarise the properties of the three functions. (The value of the angle θ has been restricted to the range of $0 \leq \theta < 360°$.)

Table 3.1

	sin θ	cos θ	tan θ
max value	1	1	Can take any value
when θ =	90°	0	positive or negative
min. value	−1	−1	Can take any value
when θ =	270°	180°	positive or negative
symmetries	Rotational	y-axis	Rotational
periodic	YES	YES	YES
period	360°	360°	180°

Transformations of the graphs

OCR P1 5.1.4 (a), 5.1.3 (g)

By making small changes in the equation, we can produce various transformations of the graphs $y = \sin x$ and $y = \cos x$, shifting it up or along, or sketching it in some direction while keeping the basic shape the same. To show the effects of these changes, we'll take $y = \sin x$ as an example, but it could equally well be $y = \cos x$.

$y = a + \sin x$

This is quite straightforward: onto the original y-value given by sin x, we're adding a, so that the original graph is shifted up by a. For example, $y = 2 + \sin x$ is the graph of $y = \sin x$ moved up 2 units:

Figure 3.19

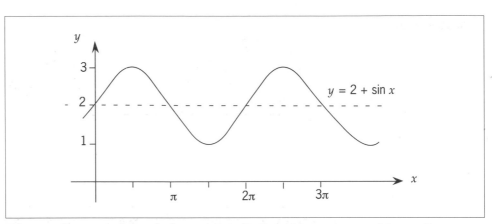

The graph passes through the y-axis at $(0, 2)$ and has a range of $1 \leq y \leq 3$: its maximum is at the same x-coordinates but the value is different. Naturally, if a happens to be negative, the graph is shifted down.

In summary: Max point $(\frac{\pi}{2}, 3)^*$ Min point $(\frac{3\pi}{2}, 1)^*$ *First positive

Range $1 \leq y \leq 3$ Period 2π

$y = \sin(x + a)$

This is more confusing: you would expect this to be a shift to the right by a, but actually it is a shift to the *left* by a, i.e. it is a translation of $-a$ in the x-direction, e.g. $y = \sin(x + \frac{\pi}{3})$ is the graph of $y = \sin x$ moved back $\frac{\pi}{3}$. If a is negative, the curve is shifted to the right.

Figure 3.20

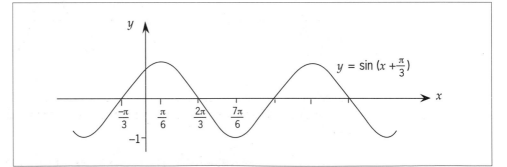

The graph passes through the y-axis before the maximum is reached – its range is still $-1 \leq y \leq 1$: the maximum has the same value but the x-coordinates of these points are different.

In summary: Max point $(\frac{\pi}{6}, 1)$ Min point $(\frac{7\pi}{6}, -1)^*$

Range $-1 \leq y \leq 1$ Period 2π

$y = a \sin x$

A scaling factor a in the y-direction, e.g. $y = 2 \sin x$ is the graph of $y = \sin x$ stretched upwards and downwards:

Figure 3.21

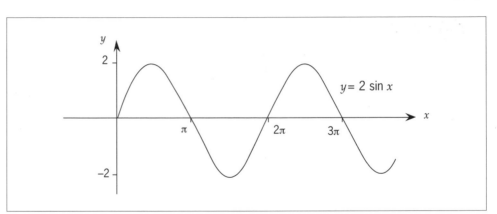

The period is unchanged, but the range has doubled: it is now $-2 \le y \le 2$.

In summary: Max point $(\frac{\pi}{2}, 2)$ Min point $(\frac{3\pi}{2}, -2)$

Range $-2 \le y \le 2$ Period 2π

$y = \sin (ax)$

A scaling, factor $\frac{1}{a}$ in the x-direction, e.g. $y = \sin (2x)$ is the graph of $y = \sin x$ squeezed in the x-direction by $\frac{1}{2}$.

Figure 3.22

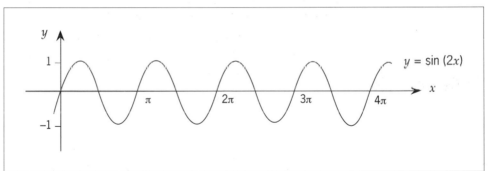

The range is still the same, but the period has now changed – it is half of what it was, i.e. it is now π.

In summary: Max point $(\frac{\pi}{4}, 1)$ Min point $(\frac{3\pi}{4}, -1)$

Range $-1 \le y \le 1$ Period π

Practice questions D

1   Sketch the graphs of the following, in each case giving the coordinates of the first maximum and minimum points, the range of y-values and the period.

   (a) $y = 1 + \cos x$ (b) $y = -1 + \sin x$

   (c) $y = \cos (x + 60°)$ (d) $y = \sin (x - \frac{\pi}{4})$

   (e) $y = 2 \cos x$ (f) $y = \frac{1}{2} \sin x$

   (g) $y = \cos 2x$ (h) $y = \sin (\frac{1}{2} x)$

Combinations of transformations

Once you're familiar with these basic transformations, you can have a go at applying a succession of them. Here's an example using three ...

| **Example** | Sketch the graph of $y = 1 + 2 \sin \left(x - \frac{\pi}{4}\right)$ for values of x between 0 and 2π, giving the coordinates of the maximum and minimum points. |

| **Solution** | We'll work from the transformations nearest the x first of all, so that in order it will be: |

$$\sin x \xrightarrow{\text{Forward } \frac{\pi}{4}} \sin \left(x - \frac{\pi}{4}\right) \xrightarrow[\text{y-direction}]{\text{Stretch}} 2 \sin \left(x - \frac{\pi}{4}\right) \xrightarrow{\text{Up } 1} 1 + 2 \sin \left(x - \frac{\pi}{4}\right)$$

The corresponding graphs will be:

Figure 3.23

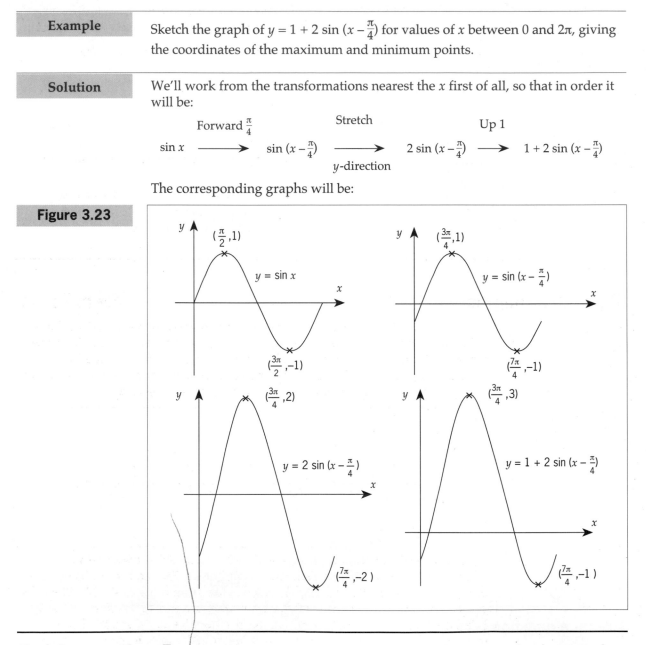

Practice questions E

1   Find the maximum and minimum values of the following expressions, together with the first value of x (≥ 0, in degrees) where they occur.

(a) $2 + \cos x$ (b) $3 - \sin x$ (c) $1 - 2 \cos x$ (d) $3 \sin 2x$

(e) $\cos (2x - 60°)$ (f) $4 + 2 \cos 3x$ (g) $1 - \sin (x + 90°)$ (h) $\sin^2 x$ (remember that $a^2 \geq 0$)

(i) $3 \cos^2 2x$ (j) $2 - 3 \sin^2 x$

2 Sketch the following graphs for $0 \leq x \leq 360°$, giving the coordinates of any maximum or minimum points:

(a) $y = -\sin x$ (Hint: this turns it upside down)

(b) $y = 1 + \cos (x + 30°)$

(c) $y = 2 + 3 \sin 2x$

(d) $y = 3 - 2 \sin (x - 60°)$

3 Sketch on the same axes for $0 \leq x \leq 360°$ the graphs of $y = \sin x$ and $y = \cos x$, and hence give the first two positive solutions of the equation $\sin x = \cos x$.

4 Find the values of the constants P and Q if $y = P + Q \sin x$ and the graph:

(a) has a range of $3 \leq y \leq 5$

(b) has a range of $-1 \leq y \leq 3$

(c) passes through the points (0,2) and $(\frac{\pi}{2}, 5)$

(d) passes through the origin and has a minimum value of –3

(e) passes through the point (30°,2) and the sum of the maximum and minimum values of y is 6.

5 Find the values of the constants P and Q if $y = P + \cos Qx$ and :

(a) the graph has a period of π and a maximum value of 3

(b) the graph has a period of 6π and minimum value of 1

(c) the graph has a minimum value of –2 when $x = 90°$

(d) the distance between the maximum and minimum values is 60° and maximum value is twice the minimum value.

Equivalent angles

OCR P1 5.1.4 (d)

Looking at a graph of $y = \sin x$, we can see that if we draw vertical lines from different points on the x-axis to meet the curve, we can find two that have the same height. In other words, the value of y, i.e. $\sin x$ can be the same for different values of x. Using the properties of periodicity and symmetry, we can find a connection between the x-values having the same y-values.

Figure 3.24

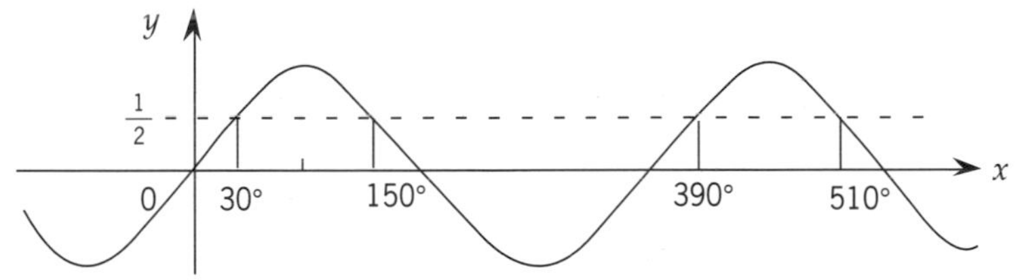

We know that the period of $\sin x$ is 360° or 2π, so $\sin 30°$ is the same as $\sin (360° + 30°) = \sin 390°$ and $\sin (360° + 390°) = \sin 750°$ and so on. There is one other angle between 0 and 360° which gives the same value as $\sin 30°$. Since the graph is symmetrical about $x = 90°$, this other angle is as much *above* 90° as 30° is *below* 90°, i.e. it is 150°. As you can see from looking at the graph, this is the same as saying that the angle is as much below 180° as 30° is above 0, which gives the same answer. We can write this as a formula:

$$\sin (180° - \theta) = \sin \theta$$

One other feature of this graph is that it is the same if you turn the page upside down, i.e. it has rotational symmetry. So, $\sin (-30°)$ is the same as $-\sin 30°$. Then, because the graph repeats exactly after 360°,

$$\sin (-30°) = \sin (-30° + 360°) = \sin (330°)$$

Also, because of the symmetry about $x = 270°$,

$$\sin (360° - 30°) = \sin (180° + 30°)$$

i.e. $\qquad \sin 330° = \sin 210°$

Figure 3.25

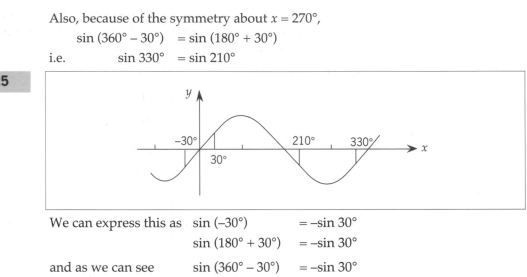

We can express this as $\quad \sin (-30°) \qquad = -\sin 30°$

$$\sin (180° + 30°) \quad = -\sin 30°$$

and as we can see $\quad \sin (360° - 30°) \quad = -\sin 30°$

and so on.

If we change the signs on both sides of the last two, we have $-\sin (180° + 30°) = \sin 30°$ and $-\sin (360° - 30°) = \sin 30°$. Collecting these together, we end up with …

$$
\begin{aligned}
\sin \theta \;&=\; \sin (180° - \theta) \\
&=\; -\sin (180° + \theta) \\
&=\; -\sin (360° - \theta) \\
&=\; \sin (360° + \theta) \\
&=\; \dots
\end{aligned}
$$

The graph of $y = \cos x$ is exactly the same as $y = \sin x$, except that it has been moved back 90° (or $\frac{\pi}{2}$ rads), so that the graph is symmetrical about $x = 0$, which means that $\cos (-\theta) = \cos \theta$. Carrying out a similar process to that for $y = \sin x$, we have the following table, written in radians for practice:

$$
\begin{aligned}
\cos \theta \;&=\; -\cos (\pi - \theta) \\
&=\; -\cos (\pi + \theta) \\
&=\; \cos (2\pi - \theta) \\
&=\; \cos (2\pi + \theta) \\
&=\; \dots
\end{aligned}
$$

and finally the least complicated one, $y = \tan x$:

$$
\begin{aligned}
\tan \theta \;&=\; -\tan (\pi - \theta) \\
&=\; \tan (\pi + \theta) \\
&=\; \dots
\end{aligned}
$$

This means that for each of the functions, there will be two angles between 0 and 360° (or 0 and 2π in radians) that give the same positive value and two angles that give the same value but negative.

For example: $\sin 30° = \dfrac{1}{2}$ $\Rightarrow$ $\sin (180° - 30°) = \dfrac{1}{2}$

i.e. $\sin 150° = \dfrac{1}{2}$

and $\sin (180° + 30°) = -\dfrac{1}{2}$

i.e. $\sin 210° = -\dfrac{1}{2}$

and $\sin (360° - 30°) = -\dfrac{1}{2}$

i.e. $\sin 330° = -\dfrac{1}{2}$

In addition, we can add 360° onto each of these angles without changing the corresponding value:

e.g. $\sin 150° = \sin (360° + 150°) = \sin 510° = \dfrac{1}{2}$

and $\sin 210° = \sin (360° + 210°) = \sin 570° = -\dfrac{1}{2}$

Practice questions F

1 If $\sin x = p$, where x is an acute angle (i.e. $0 \le x \le 90°$), find, in terms of p, the values of:

(a) $\sin (180° - x)$ (b) $\sin (360° - x)$

(c) $2 \sin (180° + x)$

(d) $\sin (180° + x) + \sin (360° + x)$

(e) $\sin (720° + x)$ (f) $\sin (540° - x)$

2 If $\cos y = q$, where y is an acute angle, find, in terms of q, the values of:

(a) $\cos (\pi - y)$ (b) $\cos (2\pi - y)$

(c) $3 - 2 \cos (\pi + y)$

(d) $\cos (3\pi - y) + \cos (3\pi + y)$

(e) $\cos (\pi + y) + \cos (2\pi + y)$

(f) $\cos (2\pi - y) - \cos (\pi - y)$

3 Find 3 additional angles in the interval $0 \le x \le 720°$ for which:

(a) $\sin x = -0.5$ (first angle is $180° + 30° = 210°$)

(b) $\cos x = 0.8$ (first angle is 36.9°)

(c) $\tan x = 4$ (first angle is 76.0°)

Give your answers to 1 d.p. where necessary.

4 If α is an acute angle, find three possible values in terms of α for θ, where $-360° \le \theta \le 360°$ and

(a) $\sin \theta = \sin \alpha$ (b) $\tan \theta = \tan \alpha$

(c) $\cos \theta = -\cos \alpha$ (d) $\sin \theta = \sin (180° + \alpha)$

(e) $\tan \theta = \tan (180° - \alpha)$

(f) $\cos \theta = \cos (360° - \alpha)$

Solving trigonometric equations

OCR P1 5.1.4 (d)

In the previous work we saw how each of the basic trigonometric functions have equivalent angles, i.e. different angles which give the same value. This means that an equation like $\sin \theta = -\dfrac{1}{2}$ has an unlimited number of solutions: a few of these are $\theta = -30°$, 210°, 330°, 570°, etc. There is quite an efficient procedure available for us to solve standard equations like these, but first of all we have to see where the various trig. ratios are positive and where they are negative.

Quadrants

There are four quadrants in a complete rotation.

Figure 3.26

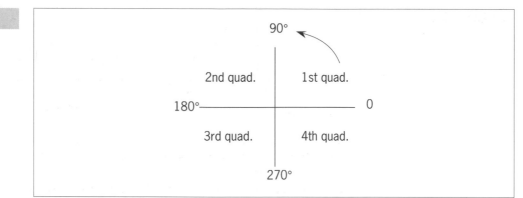

as the angle increases anticlockwise. If we look at the graphs of the trigonometric ratios, we see that sine is positive between 0 and 180°, i.e. in the first and second quadrants, cosine is positive between 0 and 90°, and 270° and 360°, i.e. in the first and fourth quadrants, and tan is positive between 0 and 90°, and 180° and 270°, in the first and third quadrants.

Gathering together the signs of the three ratios in the various quadrants we can put them in the following diagram:

Figure 3.27

②	sin +ve cos −ve tan −ve	sin +ve cos +ve tan +ve	①
③	sin −ve cos −ve tan +ve	sin −ve cos +ve tan −ve	④

We can simplify this by including only those ratios which are positive in any particular quadrant, i.e.

Figure 3.28

Sin	All
Tan	Cos

or

S	**A**
T	**C**

for short

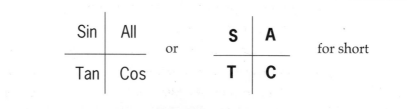

This can be remembered as '<u>A</u>ll <u>S</u>tations <u>T</u>o <u>C</u>oventry' (or similar), the initial letters as you move round anti-clockwise.

Knowing that each of the ratios is only positive in one quadrant other than the first (where they are all positive), we can see which quadrants the solutions must be in. For example, if sin θ is negative, θ must be in the Tan or the Cos quadrant, i.e. 3rd or 4th. If tan θ is positive, θ must be in the All or Tan quadrant, i.e. 1st or 3rd, etc.

Now by using symmetry, we can find out the particular solutions. If we look at the graph of sin θ

Figure 3.29

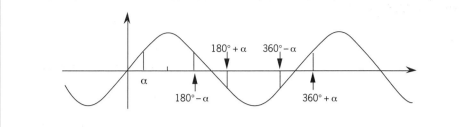

we can see that all solutions can be found once we know the value of α. If sin θ is positive, the solutions are α in the first quadrant and $180° - \alpha$ in the second. If sin θ is negative, the solutions are $180° + \alpha$ in the third and $360° - \alpha$ in the fourth. This is true for all the ratios: if we know the acute angle α (called the associated angle), solutions in the first quadrant are just α, in the second $180° - \alpha$, in the third $180° + \alpha$ and in the fourth $360° - \alpha$.

To find the value of α, we solve the original equation ignoring any minus sign. So for the equation $\sin \theta = -\frac{1}{2}$, we solve $\sin \alpha = +\frac{1}{2}$ (to distinguish from $-\frac{1}{2}$ of original equation) to give $\alpha = 30°$. For the equation $\tan \theta = \sqrt{3}$, we solve $\tan \alpha = \sqrt{3}$ to give $\alpha = 60°$.

Using this value and the symmetries of the graphs, we can find a pair of solutions in the range 0 to 360°. We then have to look at the required range given in the equation to see whether we have to adjust the solutions by adding or subtracting multiples of 360°. The range of values in the question also tells us whether the solution is to be given in degrees, for example when you are given that $0 \le \theta \le 360°$, or whether radians are indicated, e.g. $-\pi < \theta < \pi$.

In summary, the procedure is:

1 find the associated angle α

2 determine the quadrants

3 deduce a pair of solutions by symmetry

1st quadrant: $\theta = \alpha$

2nd quadrant: $\theta = (180 - \alpha)°$ or $\pi - \alpha$

3rd quadrant: $\theta = (180 + \alpha)°$ $\pi + \alpha$

4th quadrant: $\theta = (360 - \alpha)°$ $2\pi - \alpha$

4 adjust for the range if necessary.

Example

Solve $\cos \theta = -\frac{\sqrt{3}}{2}$ for $-180° < \theta < 180°$

Solution

1 The associated angle $\alpha = \text{inv cos}\left(\frac{\sqrt{3}}{2}\right) = 30°$

2 cos is negative in 2nd and 3rd quadrant

3 2nd quadrant, $180° - 30° = 150°$

3rd quadrant, $180° + 30° = 210°$

3 The angle 210° is too big – we have to subtract 360°

$\Rightarrow$ $210° - 360° = -150°$

The required solutions are 150° and –150°.

Example	Solve $\tan \theta = \dfrac{1}{\sqrt{3}}$ for values of θ in the range $0 < \theta < 2\pi$.

Solution	
1	$\alpha = $ inverse $\tan \dfrac{1}{\sqrt{3}} = 30° = \dfrac{\pi}{6}$ radians
2	tan positive 1st and 3rd quadrants
3	1st: $\theta = \alpha = \dfrac{\pi}{6}$
	3rd: $\theta = \alpha + \pi = \dfrac{7\pi}{6}$

These are in the range, so solutions are $\dfrac{\pi}{6}$ and $\dfrac{7\pi}{6}$.

Note that in equations such as these, there will be two solutions within any span of 360° (or 2π).

Example	Solve $\cos \theta = -0.71$ for $-\pi < \theta < \pi$, giving the answer to two decimal places.

Solution	
1	$\alpha = $ inverse $\cos (0.71) = 0.781$ (use at least 3 d.p. in your working)
2	cos negative in 2nd and 3rd
3	2nd: $\pi - \alpha = 2.361$
	3rd: $\pi + \alpha = 3.923$
4	The last solution is too big. Subtracting 2π gives $3.92 - 2\pi = -2.361$.

Solutions are -2.36 and 2.36 (to 2 d.p.)

Practice questions G

1 Solve the following equations for $0 \le \theta \le 360°$:

(a) $\sin \theta = \dfrac{1}{2}$
(b) $\cos \theta = \dfrac{1}{\sqrt{2}}$
(c) $\tan \theta = -\sqrt{3}$
(d) $\sin \theta = -0.2$
(e) $\cos \theta = -0.4$
(f) $\tan \theta = 1.7$
(g) $5 \sin \theta = 3$
(h) $3 \cos \theta - 2 = 0$
(i) $\dfrac{1}{\tan \theta} = 4.1$
(j) $2 \sin \theta + 2 = 1.7$

2 Solve the following equations for $0 \le \theta \le 2\pi$, giving your answer in terms of π. (Find the angle first and then convert.)

(a) $\sin \theta = -\dfrac{1}{2}$
(b) $\cos \theta = \dfrac{\sqrt{3}}{2}$
(c) $\tan \theta = -1$
(d) $\sin \theta = 1$
(e) $\cos \theta = 0$
(f) $\sqrt{3} \tan \theta = 1$
(g) $\sqrt{2} \sin \theta = -1$
(h) $2 \cos \theta + 1 = 0$
(i) $\sqrt{3} \tan \theta = -3$
(j) $2 \sin \theta - \sqrt{3} = 0$

3 Solve the following equations for $-180° \le \theta \le 180°$, giving your answers correct to 1 d.p. where necessary:

(a) $\sin \theta = -0.7$
(b) $\cos \theta = 0.8$
(c) $\tan \theta = 5$
(d) $3 \sin \theta - 1 = 0$
(e) $5 \cos \theta + 2 = 0$
(f) $3 (1 + \tan \theta) = 2$

4 Solve the following equations for $0 \le \theta \le 2\pi$, giving your answer to 2 d.p. Make sure your calculator is in radians.

(a) $\sin \theta = -0.81$
(b) $\cos \theta = 0.45$
(c) $\tan \theta = 1.8$
(d) $2 \sin \theta = 0.7$
(e) $4 \cos \theta - 3 = 0$
(f) $2 \tan \theta + 7 = 2.4$

Further equations

We often have to solve equations such as $\sin 2x = \dfrac{\sqrt{3}}{2}$ or $\tan(x + 50°) = 1$, where the argument is more complicated than just x. We can make a simple substitution, in the given cases $y = 2x$ or $y = x + 50°$, solve the equation for y and deduce the appropriate values for x.

There is a corresponding change in the range when we change the variable, so we have to be careful to include all the possible solutions (and exclude any outside the range).

Example	Solve the equation $\sin 2x = \dfrac{\sqrt{3}}{2}$ for $0 \le x \le 360°$.

Solution	Putting $y = 2x$ changes the equation into

$$\sin y = \frac{\sqrt{3}}{2} \quad \text{for } 0 \le \frac{y}{2} \le 360° \text{ (since } x = \frac{y}{2}\text{)}, \text{ i.e. } 0 \le y \le 720° \text{ (multiplying by 2)}$$

The associated angle is 60°, positive in the 1st and 2nd quadrants, so 60° and $180° - 60° = 120°$ are two solutions. but we can now add 360° onto both of these and still stay within the range (for y). So now we have four solutions, which are:

$$y = 60°, 120°, 420° \text{ and } 480°.$$

Substituting back $y = 2x$,

$$2x = 60°, 120°, 420° \text{ and } 480° \qquad \text{i.e. } x = 30°, 60°, 210°, 240°$$

which you can see are all within the given limits.

Example	Solve the equation $\tan(x + 50°) = 1$ for $0 \le x \le 360°$.

Solution	Putting $y = x + 50°$ into the equation gives

$$\tan y = 1 \quad \text{for } 0 \le y - 50° \le 360°$$
$$\text{i.e. } 50° \le y \le 410° \text{ (adding 50°)}$$

$\alpha = 45°$, positive in 1st and 3rd, $y = 45°$ or $225°$.

The first of these is too small, so adding 360° gives the two solutions

$$y = 225° \text{ or } 45° + 360° = 405°$$
$$\Rightarrow \quad x + 50° = 225° \text{ or } 405°$$
$$\Rightarrow \quad x = 175° \text{ or } 355°$$

Example	Solve $\cos(2x - 0.7) = -0.8$, within the range $-\pi < x < \pi$.

Solution	Putting $y = 2x - 0.7$ into the equation gives

$$y + 0.7 = 2x \quad \Rightarrow x = \frac{y + 0.7}{2}$$

The equation becomes $\cos y = -0.8$ for $-\pi < \dfrac{y + 0.7}{2} < \pi$

i.e. $-2\pi < y + 0.7 < 2\pi$

i.e. $-6.983 < y < 5.583$

Associated angle is 0.644, negative in 2nd and 3rd,

so $\pi - 0.644 = 2.498$

$\pi + 0.644 = 3.785$

We can't *add* 2π to either of these without exceeding the limits, but we can subtract

$2.498 - 2\pi = -3.785$

$3.785 - 2\pi = -2.498$

$\Rightarrow \quad 2x - 0.7 = 2.498 \quad \Rightarrow \quad x = 1.60$

$2x - 0.7 = 3.785 \quad \Rightarrow \quad x = 2.24$

and the same but negative, i.e.

$x = \pm\,1.60$ or $\pm\,2.24$

Practice questions H

1 Find the values of θ between 0 and 360° such that:

(a) $\sin 2\theta = 0.5$

(b) $\cos 2\theta = -0.5$

(c) $\tan 2\theta = -1$

giving your answers in degrees to 1 d.p. (where necessary)

2 Solve the following equations for $-180° \le \theta \le 180°$:

(a) $\cos(x + 30°) = \dfrac{1}{\sqrt{2}}$

(b) $\sin(x - 60°) = \dfrac{\sqrt{3}}{2}$

(c) $\tan(x + 10°) = 0.8$

giving your answers to 1 d.p. (where necessary)

3 Solve the following equations for $0 \le x \le 2\pi$, giving your answers in radians to 2 d.p.

(a) $\sin(2x - \dfrac{\pi}{2}) = 0.7$

(b) $\tan(2x + \dfrac{\pi}{3}) = -3$

(c) $\cos(2x - 0.5) = \dfrac{\sqrt{3}}{2}$

4 Solve the following equations, giving your answers to 3 s.f.

(a) $\sin(x + 20°) = 0.9$ $\qquad 180° \le x \le 540°$

(b) $\cos 3x = \dfrac{1}{3}$ $\qquad 0 < x < 270°$

(c) $\tan 2x = 1$ $\qquad \dfrac{\pi}{2} < x < \dfrac{3\pi}{2}$

(d) $\cos(2x + \dfrac{\pi}{6}) = 1$ $\qquad (-\pi < x < \pi)$

Quadratic equations

OCR P1 5.1.4 (d)

The simplest of these is of the type

$$\sin^2 \theta = \tfrac{1}{2} \quad \text{for } 0 \le \theta \le 2\pi$$

Here we have to remember, when we take square roots of both sides, that a negative number squared is also positive, so we have the two possibilities:

$$\sin \theta = +\frac{1}{\sqrt{2}} \qquad \text{or} \qquad \sin \theta = -\frac{1}{\sqrt{2}}$$

The associated angle is $\dfrac{\pi}{4}$, so the first equation gives $\dfrac{\pi}{4}$ and $\pi - \dfrac{\pi}{4} = \dfrac{3\pi}{4}$.

The second gives $\pi + \dfrac{\pi}{4} = \dfrac{5\pi}{4}$ and $2\pi - \dfrac{\pi}{4} = \dfrac{7\pi}{4}$.

Altogether,

$$\theta = \frac{\pi}{4}, \frac{3\pi}{4}, \frac{5\pi}{4} \text{ or } \frac{7\pi}{4}$$

We can also find equations of the type

$$a \sin^2 \theta + b \sin \theta + c = 0$$

which is the general form of a quadratic equation, but with the familiar x replaced by $\sin \theta$ (or one of the other ratios, $\cos \theta$ or $\tan \theta$). The method of solution is the same: factorise if we can and otherwise use the formula. This can give values of $\sin \theta$ or $\cos \theta$ that are not possible – these solutions are simply rejected.

| **Example** | Solve the equation |

$$2 \sin^2 \theta + 3 \sin \theta - 2 = 0 \qquad \text{for } 0 \le \theta \le 360°$$

| **Solution** | This factorises quite nicely: |

$$2 \sin^2 \theta + 3 \sin \theta - 2 = (2 \sin \theta - 1)(\sin \theta + 2) = 0$$

$$\Rightarrow \quad \sin \theta = \frac{1}{2}, \text{ so } \theta = 30° \text{ or } 150°$$

or $\quad \sin \theta = -2$, which is impossible.

The only solutions are $\theta = 30°$ or $150°$.

| **Example** | Solve the equation |

$$3 \tan^2 \theta - 11 \tan \theta + 5 = 0$$

| **Solution** | This does not factorise, so we have to use the quadratic formula: |

$$\tan \theta = \frac{11 \pm \sqrt{11^2 - 4(3)(5)}}{6}$$

$$= \frac{11 \pm \sqrt{51}}{6} = 3.024 \text{ or } 0.643$$

$$\tan \theta = 3.024 \quad \Rightarrow \quad \theta = 71.7° \text{ or } 251.7°$$

$$\tan \theta = 0.643 \quad \Rightarrow \quad \theta = 32.7° \text{ or } 212.7°$$

Note that there are no rejected solutions for tan.

Practice questions I

1 Solve the following equations for $-\pi \le \theta \le \pi$, giving your answers to 2 d.p.

(a) $\tan^2 \theta = 4$

(b) $\sin^2 \theta = \frac{1}{3}$

(c) $4 \cos^2 \theta = 1$

(d) $3 + 8 \sin^2 \theta = 5.88$

(e) $(1 + \cos \theta)^2 = 2$

(f) $\tan^3 \theta = 2 \tan \theta$

(g) $\dfrac{2}{\cos \theta} = 3 \cos \theta$

2 Solve the following equations for $0 \le \theta \le 360°$:

(a) $2 \cos^2 \theta + \cos \theta - 1 = 0$

(b) $3 \sin^2 \theta + 5 \sin \theta - 4 = 0$

(c) $2 \sin^2 \theta = \sin \theta$

(d) $\tan^2 \theta - 6 \tan \theta = 9$

(e) $6 \cos^2 \theta + \cos \theta = 2$

Pythagorean identity

OCR P1 5.1.4 (b)

Equations also occur with mixtures of ratios, particularly $\sin \theta$ and $\cos \theta$. We cannot use the quadratic formula until the equation has been changed to one in which there is only one variable. To do this, we need a formula that connects the values of $\sin \theta$ and $\cos \theta$. This comes from looking at the ratios of sides in a right-angled triangle and using Pythagoras' Theorem.

Figure 3.30

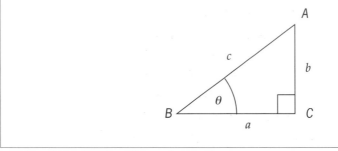

In this triangle, $\sin \theta = \dfrac{b}{c}$ and $\cos \theta = \dfrac{a}{c}$,

so $\quad \sin^2 \theta + \cos^2 \theta = \dfrac{b^2}{c^2} + \dfrac{a^2}{c^2} = \dfrac{b^2 + a^2}{c^2}$

But by Pythagoras' Theorem, $b^2 + a^2 = c^2$,

so $\quad \sin^2 \theta + \cos^2 \theta = \dfrac{c^2}{c^2} = 1$

This is a very important identity:

$$\sin^2 \theta + \cos^2 \theta = 1$$

It has two immediate uses:

1 It allows us to find the value for $\sin \theta$ knowing the value of $\cos \theta$, and the other way round.

2 In the form $\sin^2 \theta = 1 - \cos^2 \theta$, we can substitute into an equation with $\sin^2 \theta$ and $\cos^2 \theta$ to make a quadratic equation in a single variable.

Example

Find the value of $\cos \theta$, given that $\sin \theta = 0.8$ and $90° < \theta < 180°$.

Solution

Using $\sin^2 \theta + \cos^2 \theta = 1$ and $\sin \theta = 0.8$ gives

$$(0.8)^2 + \cos^2 \theta = 1$$
$$\Rightarrow \quad \cos^2 \theta = 0.36$$
$$\Rightarrow \quad \cos \theta = \pm 0.6$$

Since θ is in the second quadrant, where $\cos \theta$ is negative, we choose the negative solution:

$$\cos \theta = -0.6$$

Example	Solve the equation

$$2 \sin^2 \theta + 5 \cos \theta + 1 = 0 \qquad \text{for } -\pi < \theta < \pi$$

giving your answers exactly in terms of π.

Solution	Using $\sin^2 \theta = 1 - \cos^2 \theta$, the equation becomes

$$2(1 - \cos^2 \theta) + 5 \cos \theta + 1 = 0$$
$$2 - 2 \cos^2 \theta + 5 \cos \theta + 1 = 0$$
$$3 - 2 \cos^2 \theta + 5 \cos \theta = 0$$
$$2 \cos^2 \theta - 5 \cos \theta - 3 = 0$$
$$(\cos \theta - 3)(2 \cos \theta + 1) = 0$$
$$\cos \theta = 3 \text{ rejected}$$

or $\quad \cos \theta = -\dfrac{1}{2} \implies \theta = 120° \text{ or } 240°$

$$= \frac{2\pi}{3} \text{ or } \frac{4\pi}{3}.$$

$\dfrac{4\pi}{3}$ is too big, so $\dfrac{4\pi}{3} - 2\pi = \dfrac{-2\pi}{3}$

so $\quad \theta = \dfrac{2\pi}{3}$ or $\theta = \dfrac{-2\pi}{3}$ are the solutions.

Practice questions J

1 (a) Find the exact value of $\cos \theta$ if $\sin \theta = \dfrac{12}{13}$
and $270° < \theta < 360°$.

   (b) Find the value of $\sin \theta$ in the form $\dfrac{a}{b}\sqrt{7}$
where a and b are integers, if θ is an acute
angle and $\cos \theta = \dfrac{1}{8}$

2 Solve the following equations for $0 < \theta < 360°$,
giving your answers in degrees to 1 d.p.

   (a) $3 \cos^2 \theta - \sin \theta - 1 = 0$

   (b) $2 \sin^2 \theta + \cos \theta = 1$

   (c) $6 \sin^2 \theta + 5 \cos \theta = 0$

   (d) $5 \cos \theta = 2 + 3 \sin^2 \theta$

   (e) $2 - \sin \theta = \cos^2 \theta + 7 \sin^2 \theta$

Equations involving tan θ

OCR P1 5.1.4 (b)

If we look back at the right-angled triangle in Fig. 3.30, we see that $\tan \theta = \dfrac{b}{a}$.

As we saw, $\sin \theta = \dfrac{b}{c}$ and $\cos \theta = \dfrac{a}{c}$, and so

$$\frac{\sin \theta}{\cos \theta} = \frac{b}{c} \div \frac{a}{c} = \frac{b}{c} \times \frac{c}{a} = \frac{b}{a} \text{ also.}$$

This is true for any angle θ.

$$\tan \theta = \frac{\sin \theta}{\cos \theta}$$

This is also a very useful identity and allows us to solve many further equations.

| **Example** | Solve $2 \cos \theta = \sin \theta$ for $-180° < \theta < 180°$ |

| **Solution** | We can divide both sides by $\cos \theta$ to give |

$$2 = \frac{\sin \theta}{\cos \theta} = \tan \theta$$

$$\tan \theta = 2 \implies \theta = 63.4° \text{ or } 243.4°$$

| **Example** | Solve $\tan \theta = 2 \sin \theta$ for $0° < \theta < 360°$ |

| **Solution** | Put $\tan \theta = \dfrac{\sin \theta}{\cos \theta}$ and the equation becomes |

$$\frac{\sin \theta}{\cos \theta} = 2 \sin \theta$$

Here we can't divide through by $\sin \theta$: this loses solutions. Multiply through by $\cos \theta$ and rearrange, so

$$\sin \theta = 2 \sin \theta \cos \theta$$

$$2 \sin \theta \cos \theta - \sin \theta = 0$$

$$\sin \theta (2 \cos \theta - 1) = 0$$

$$\sin \theta = 0 \qquad \implies \theta = 0°, 180°, 360°$$

$$2 \cos \theta - 1 = 0 \implies \cos \theta = \frac{1}{2}, \ \theta = 60° \text{ or } 300°$$

So the solutions are 0, 60°, 180°, 300°, 360°

| **Example** | Solve the equation |

$$6 \cos \theta = 5 \tan \theta \qquad \text{for } 0 < \theta < 360°$$

giving answers to the nearest degree.

| **Solution** | Putting $\tan \theta = \dfrac{\sin \theta}{\cos \theta}$, the equation becomes |

$$6 \cos \theta = \frac{5 \sin \theta}{\cos \theta}$$

$$6 \cos^2 \theta = 5 \sin \theta \qquad (\text{use } \cos^2 \theta = 1 - \sin^2 \theta)$$

$$6 (1 - \sin^2 \theta) = 5 \sin \theta$$

$$6 - 6 \sin^2 \theta = 5 \sin \theta$$

$$6 \sin^2 \theta + 5 \sin \theta - 6 = 0$$

$$(3 \sin \theta - 2)(2 \sin \theta + 3) = 0$$

$$\implies \sin \theta = \frac{2}{3}, \quad \theta = 42° \text{ or } 138°$$

or $\sin \theta = -\dfrac{3}{2}$ rejected

Practice questions K

1 Solve the following equations for $-180° < \theta < 180°$, giving your answers in degrees to 1 d.p., where appropriate:

(a) $\cos \theta = 3 \sin \theta$ (b) $2 \sin \theta + \cos \theta = 0$ (c) $\sin \theta (\sin \theta - \cos \theta) = 0$ (d) $3 \tan \theta = 4 \sin \theta$

(e) $\tan \theta + 3 \sin \theta = 0$ (f) $3 \tan \theta = 2 \cos \theta$ (g) $3 \sin^2 \theta = \cos^2 \theta$ (h) $\sin 3\theta = 3 \cos 3\theta$

SUMMARY EXERCISE

1

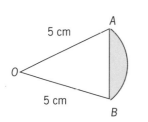

Figure 1 shows a sector OAB of a circle, centre O, of radius 5 cm and a shaded segment of the circle. Given that $\angle AOB = 0.7$ radians, calculate

(a) the area, in cm², of the sector OAB.

(b) the area, in cm² to 2 significant figures, of the shaded segment.

2

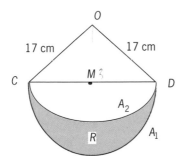

The diagram shows the triangle OCD with $OC = OD = 17$ cm and $CD = 30$ cm. The mid-point of CD is M. With centre M, a semicircular arc A_1 is drawn on CD as diameter. With centre O and radius 17 cm, a circular arc A is drawn from C to D. The shaded region R is bounded by the arcs A_1 and A_2. Calculate, giving answers to 2 decimal places,

(a) the area of the triangle OCD

(b) the angle COD in radians

(c) the area of the shaded region R.

3 Two circles with centres A and B, each of radius 13 cm, lie in a plane with their centres 24 cm apart. The circles intersect at the points C and D.

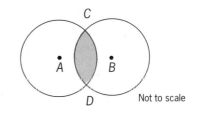

Not to scale

(a) Determine the length of CD.

(b) Calculate the size of angle CAD in radians, giving your answer to four significant figures

(c) Calculate the area of the shaded region common to both circles, giving your answer to the nearest 0.1 cm².   [AEB 1993]

4 On a circular clock face, with centre O, the minute hand OA is of length 10 cm and the hour hand OB is of length 6 cm. Prove that the difference between the distances that A and B travel during the period of 1 hour from 12 o'clock to 1 o'clock is 19π cm.

5 Find all values of x for which $0° < x < 360°$ that satisfy the equation

$$\sin \left(\tfrac{1}{2} x \right) = \tfrac{1}{4}.$$

6 Find all solutions of the equation

$$2 \cos^2 x + 3 \sin x = 0$$

in the interval $0° \le x \le 360°$.   [AEB 1995]

7 Find all solutions in the interval $0° < \theta < 180°$ of the equation

$$2 \sin (\theta - 48°) - 1 = 0$$ [AEB 1997]

8 Find, to the nearest integer, the values of x in the interval $0 \le x < 180$ for which

$$3 \sin^2 3x° - 7 \cos 3x° - 5 = 0.$$

9 Find, giving your answers in terms of π, all values of θ in the interval $0 \le \theta < 2\pi$ for which

(a) $\tan\left(\theta + \frac{\pi}{3}\right) = 1$,

(b) $\sin 2\theta = -\frac{\sqrt{3}}{2}$.

10 (a) Given that $\tan 75° = 2 + \sqrt{3}$, find in the form $m + n\sqrt{3}$, where m and n are integers, the value of

　(i)　$\tan 15°$

　(ii)　$\tan 105°$.

(b) Find, in radians to two decimal places, the values of x in the interval $0 \le x \le 2\pi$, for which $3 \sin^2 x + \sin x - 2 = 0$.

11 Given that A is the obtuse angle such that $\sin A = \frac{1}{3}$, find the value of $\cos A$, expressing your answer in surd form.

12 The depth of water at the end of a pier is H metres at time t hours after high tide, where

$$H = 20 + 5 \cos (kt)$$

and k is a positive constant measured in radians per hour.

(a) Write down the greatest and least depths of water at the end of the pier.

(b) Sketch the graph of H against t from $t = 0$ to the time of the next high tide.

(c) The time interval between successive high tides is 11 hours and 15 minutes. Calculate the value of k, giving your answer to three significant figures.

13

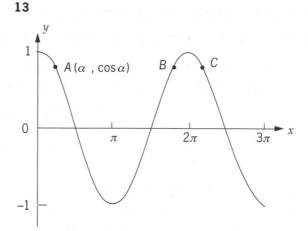

The diagram shows the graph of $y = \cos x$ for $0 \le x \le 3\pi$. The point A has coordinates again at B and C.

(a) Write down the x-coordinates of B and C.

(b) Solve the inequality $\cos x > \cos \alpha$, for $0 \le x \le 3\pi$.

14 $f(x) = 3 + 2 \sin (2x + k)°$, $0 \le x < 360$, where k is a constant and $0 < k < 360$. The curve with equation $y = f(x)$ passes through the point with coordinates $(15, 3 + \sqrt{3})$.

(a) Show that $k = 30$ is a possible value for k and find the other possible value of k.

Given that $k = 30$

(b) solve the equation $f(x) = 1$

(c) find the range of f.

(d) sketch the graph of $y = f(x)$, stating the coordinates of the turning points and the coordinates of the point where the curve meets the y-axis.

SUMMARY

This section has been an introduction to a very important group of functions, their graphs and related equations. Make sure you are able to change from radians to degrees and the other way round – and remember to change the mode in your calculator accordingly.

Having finished this chapter, you should now:

● be familiar with radian measure and how this relates to degrees

● be able to change from degrees to radians and vice versa

● know the formulas for arcs and sectors: $s = r\theta$ and $A = \frac{1}{2} r^2 \theta$

- know the difference between an arc and a chord
- be familiar with the formula for the area of a triangle: $\Delta = \frac{1}{2} ab \sin C$ ✓
- be able to calculate the perimeter and area for the regions related to circles
- be familiar with the graphs of sin θ, cos θ and tan θ.
- know when the maximum and minimum values of these occur (if any) and what the values are
- know the period and symmetries of these graphs and be able to use these to deduce equivalent angles
- know the effect of transforming the equation on the corresponding graph
- be able to combine these transformations in the correct order
- know the procedure for solving trig. equations
- be able to apply it to equations such as $\sin \theta = a$
- know when to use radians and when to use degrees
- be able to give solutions within a specified range
- know how to solve equations of the type $\sin a\theta = b$ and $\cos (\theta + c) = d$, including adjusting the range
- know that if $\cos^2 \theta = a$, cos θ has the two possible values $\sqrt{a}$ and $-\sqrt{a}$
- be able to solve equations of the type $a \tan^2 \theta + b \tan \theta + c = 0$
- know that $\sin^2 \theta + \cos^2 \theta = 1$
- be able to use this to convert an equation in two ratios into an equation in just one
- know that $\tan \theta - \dfrac{\sin \theta}{\cos \theta}$ and use this identity to solve equations.

ANSWERS

Practice questions A

1   (a) 2π (b) $\frac{\pi}{3}$ (c) $\frac{3\pi}{2}$ (d) 3π

     (e) $\frac{2\pi}{3}$ (f) $\frac{4\pi}{3}$ (g) $\frac{5\pi}{12}$ (h) $\frac{5\pi}{4}$

     (i) $\frac{7\pi}{4}$ (j) $\frac{3\pi}{4}$

2   (a) 60°   (b) 135°   (c) 120°   (d) 150°

     (e) 270°   (f) 720°   (g) 225°   (h) 210°

     (i) 300°   (j) 18°

Practice questions B

1   (a) $2\pi, 4\pi$ (b) $14\pi, 84\pi$

     (c) $6\pi, 24\pi$ (d) $15\pi, 67.5\pi$

2   6

3   (a) 7   (b) $\frac{147}{2}$

Practice questions C

1   (a) 21.4   (b) 134.2   (c) 27.3

2   (a) 28.3   (b) 942.5   (c) 24.3

3   (a) 1.77   (b) 1.58   (c) 3.96   (d) 4.87

4   2.86 cm^2

Practice questions D

1 (a)

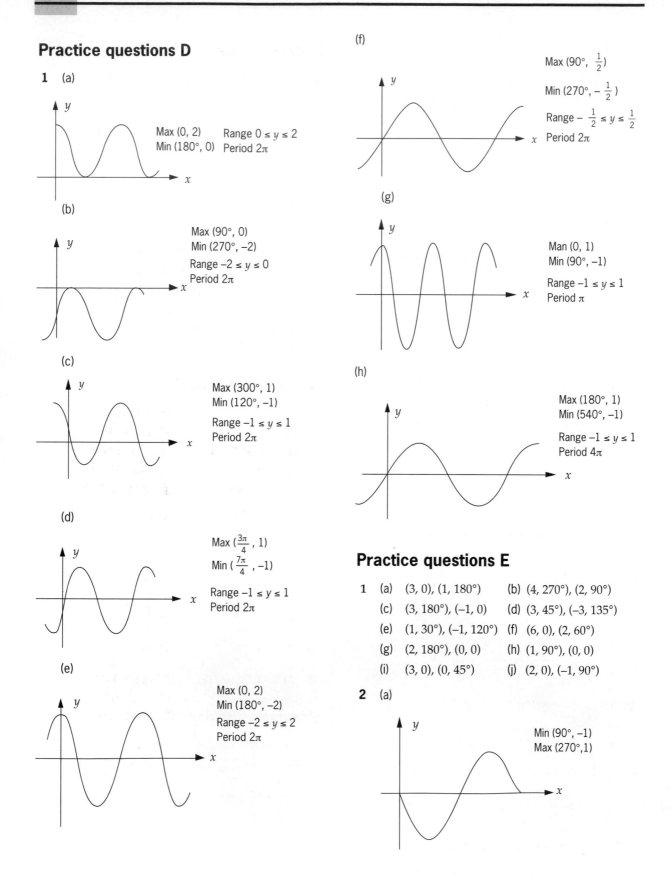

Max (0, 2)   Range $0 \leq y \leq 2$
Min (180°, 0)   Period 2π

(b)

Max (90°, 0)
Min (270°, −2)
Range $-2 \leq y \leq 0$
Period 2π

(c)

Max (300°, 1)
Min (120°, −1)
Range $-1 \leq y \leq 1$
Period 2π

(d)

Max $(\frac{3\pi}{4}, 1)$
Min $(\frac{7\pi}{4}, -1)$
Range $-1 \leq y \leq 1$
Period 2π

(e)

Max (0, 2)
Min (180°, −2)
Range $-2 \leq y \leq 2$
Period 2π

(f)

Max (90°, $\frac{1}{2}$)
Min (270°, $-\frac{1}{2}$)
Range $-\frac{1}{2} \leq y \leq \frac{1}{2}$
Period 2π

(g)

Man (0, 1)
Min (90°, −1)
Range $-1 \leq y \leq 1$
Period π

(h)

Max (180°, 1)
Min (540°, −1)
Range $-1 \leq y \leq 1$
Period 4π

Practice questions E

1 (a)   (3, 0), (1, 180°)   (b)   (4, 270°), (2, 90°)
(c)   (3, 180°), (−1, 0)   (d)   (3, 45°), (−3, 135°)
(e)   (1, 30°), (−1, 120°)   (f)   (6, 0), (2, 60°)
(g)   (2, 180°), (0, 0)   (h)   (1, 90°), (0, 0)
(i)   (3, 0), (0, 45°)   (j)   (2, 0), (−1, 90°)

2 (a)

Min (90°, −1)
Max (270°, 1)

(b)

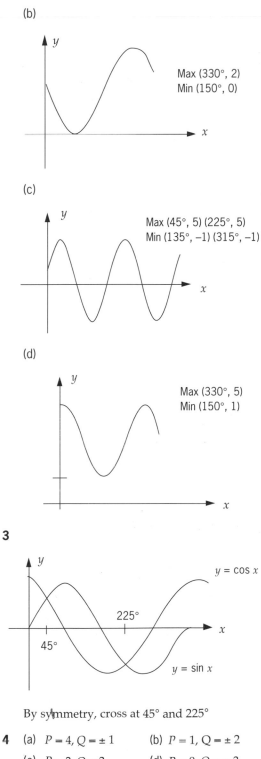

Max (330°, 2)
Min (150°, 0)

(c)

Max (45°, 5) (225°, 5)
Min (135°, –1) (315°, –1)

(d)

Max (330°, 5)
Min (150°, 1)

3

$y = \cos x$

225°

45°

$y = \sin x$

By symmetry, cross at 45° and 225°

4 (a) $P = 4, Q = \pm 1$ (b) $P = 1, Q = \pm 2$
(c) $P = 2, Q = 3$ (d) $P = 0, Q = \pm 3$
(e) $P = 3, Q = -2$

5 (a) $P = 2, Q = 2$ (b) $P = 2, Q = \frac{1}{3}$
(c) $P = -1, Q = 2$ (d) $P = 3, Q = 3$

Practice questions F

1 (a) p (b) $-p$ (c) $-2p$ (d) 0
(e) p (f) p

2 (a) $-q$ (b) q (c) $3 + 2q$ (d) $-2q$
(e) 0   (f) $2q$

3 (a) 330°, 570°, 690°
(b) 323.1°, 396.9°, 683.1°
(c) 256.0°, 436.0°, 616.0°

4 (a) $-360 + \alpha, -180 - \alpha, 180 - \alpha$
(b) $-360 + \alpha, -180 + \alpha, 180 + \alpha$
(c) $180 - \alpha, 180 + \alpha, -180 + \alpha$
(d) $-\alpha, -180 + \alpha, 180 + \alpha$
(e) $-\alpha, 180 - \alpha, 360 - \alpha$
(f) $\alpha, 360 - \alpha, -\alpha$

Practice questions G

1 (a) 30°, 150°   (b) 45°, 315°
(c) 120°, 300°   (d) 191.5°, 348.5°
(e) 113.6°, 246.4°   (f) 59.5°, 239.5°
(g) 36.9°, 143.1°   (h) 48.2°, 311.8°
(i) 13.7°, 193.7°   (j) 188.6°, 351.4°

2 (a) $\frac{7\pi}{6}, \frac{11\pi}{6}$ (b) $\frac{\pi}{6}, \frac{11\pi}{6}$
(c) $\frac{3\pi}{4}, \frac{7\pi}{4}$ (d) $\frac{\pi}{2}$
(e) $\frac{\pi}{2}, \frac{3\pi}{2}$ (f) $\frac{\pi}{6}, \frac{7\pi}{6}$
(g) $\frac{5\pi}{4}, \frac{7\pi}{4}$ (h) $\frac{2\pi}{3}, \frac{4\pi}{3}$
(i) $\frac{2\pi}{3}, \frac{5\pi}{3}$ (j) $\frac{\pi}{3}, \frac{2\pi}{3}$

3 (a) $-44.4°, -135.6°$ (b) $-36.9, 36.9°$
(c) $-101.3°, 78.7°$ (d) $19.5°, 160.5°$
(e) $-113.6°, 113.6°$ (f) $-18.4°, 161.6°$

4 (a) 4.09, 5.34   (b) 1.10, 5.18
(c) 1.06, 4.21   (d) 0.36, 2.78
(e) 0.72, 5.56   (f) 1.98, 5.12

Practice questions H

1 (a) 15°, 75°, 195°, 255°

(b) 60°, 120°, 240°, 300°

(c) $67\frac{1}{2}°$, $157\frac{1}{2}°$, $247\frac{1}{2}°$, $337\frac{1}{2}°$

2 (a) −75°, 15°

(b) 120°, 180°

(c) 28.7°, −151.3

3 (a) 1.17, 1.97, 4.31, 5.11

(b) 0.42, 1.99, 3.56, 5.14

(c) 0.51, 3.13, 3.65, 6.27

4 (a) 404°, 456°

(b) 23.5°, 96.5°, 143.5°, 216.5°, 263.5°

(c) $\dfrac{5\pi}{8}$, $\dfrac{9\pi}{8}$

(d) $\dfrac{-\pi}{12}$, $\dfrac{11\pi}{12}$

Practice questions I

1 (a) −1.11, −2.03, 1.11, 2.03

(b) ± 0.62, ±2.53

(c) ± 1.05, ± 5.24

(d) ± 0.64, ± 2.50

(e) ± 1.14

(f) ± 3.14, 0, ± 0.96, ± 2.19

(g) ± 0.62, ± 2.53

2 (a) 60°, 180°, 300°

(b) 36.2°, 143.8°

(c) 0°, 30°, 150°, 180°, 360°

(d) 82.1°, 128.8°, 262.1°, 308.8°

(e) 60°, 131.8°, 228.2°, 300°

Practice questions J

1 (a) $\dfrac{5}{13}$ (b) $\dfrac{3}{8}\sqrt{7}$

2 (a) 41.8°, 138.2° (b) 120°, 240°

(c) 131.8°, 228.2° (d) 45.3°, 314.7°

(e) 19.5°, 160.5°, 210°, 330°

Practice questions K

1 (a) −161.6, 18.4 (b) −26.6, 153.4

(c) 0, −135°, 45° (d) ±41.4°, 0

(e) ±109.5°, 0 (f) 30°, 150°

(g) ±30°, ±150°

(h) 23.9, 83.9, 143.9°, −156.1°, −96.1°, −36.1°

Coordinate geometry

The simplest relationship between two variables is a linear one. If y increases by a certain amount when x increases by one, then double the increase in x means double the increase in y. We can recognise this relationship very quickly when we see the graphical equivalent: it is a straight line. In this section we look at the different types of equation that occur within this particular family and how we can devise these equations given certain conditions.

The gradient of a line

OCR P1 5.1.3 (a)

Equations involving only single powers of x and y, like $y = 3x - 2$ or $2x + y + 8 = 0$, are called *linear equations* because their corresponding graphs are straight lines. We can find the value of the variable y which corresponds to a given value of x simply by substituting the x-value into the equation and solving. If we take the first equation $y = 3x - 2$, and substitute $x - 0$ we find that $y = 3 \times 0 - 2 = -2$. Similarly when $x = 1$, $y = 1$ and when $x = 2$, $y = 4$. If we plot these points and join them, we end up with the following line:

Figure 4.1

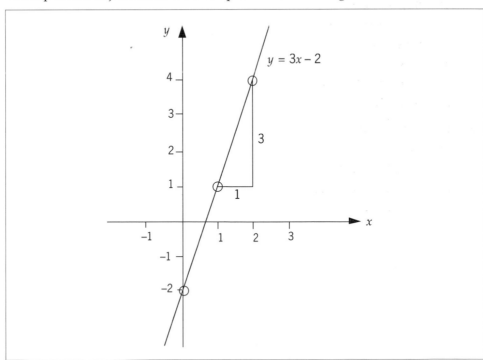

It is quite a steep line – as x increases from 1 to 2, y increases from 1 to 4. The ratio between y-difference and x-difference is called the *gradient*, so that

$$\text{gradient} = \frac{y\text{-increase}}{x\text{-increase}}$$

The gradient for the line is then $\frac{3}{1} = 3$.

As you may already know, we could have found the gradient of this line directly by looking at its equation, provided this is in the form $y = mx + c$, and seeing the coefficient of the x-term. In the same way, by looking at the equation $y = -4x + 2$ we can say that it is a straight line with gradient -4. A negative gradient means that y *decreases* so much for every increase of 1 by x, so that the line slopes the other way.

Let's take two points from this equation, say $x = 0$, $y = 2$ and $x = 1$, $y = -2$, plot these and join them.

Figure 4.2

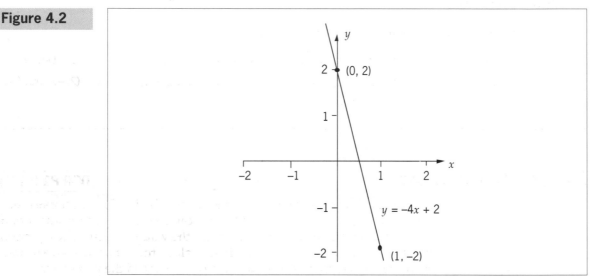

You can see that as x increases from 0 to 1, y decreases from 2 to -2, so that the gradient is $\dfrac{y\text{-increase}}{x\text{-increase}} = \dfrac{-4}{+1} = -4$, as we had already seen.

In general, if (x_1, y_1) and (x_2, y_2) are any two points on the line:

Figure 4.3

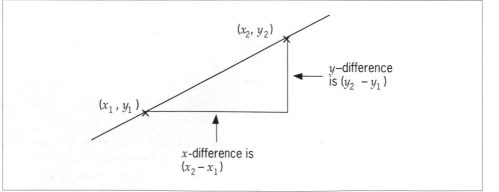

we find the gradient using the definition:

$$\text{gradient} = \frac{y\text{-difference}}{x\text{-difference}} = \frac{y_2 - y_1}{x_2 - x_1}$$

This is a very useful result.

If (x_1, y_1) and (x_2, y_2) are any two points,

the gradient m of the line joining them is given by

$$m = \frac{y_2 - y_1}{x_2 - x_1}$$

Practice questions A

Find the gradient of the line joining the following pairs of points

1 (a) (3, 2) and (5, 6)   (b) (–3, 1) and (4, 8)   (c) (0, 0) and (2, 4)   (d) (1, –1) and (–1, 1)

    (e) (2, 2) and (4, –4)   (f) (–1, 2) and (3, 4)   (g) (–1, –3) and (–2, 5)   (h) (2, –3) and (–1, 4)

    (i) (2, 5) and (4, 5)   (j) (3, 2) and (3, 5)

The equation of a line   **OCR P1** 5.1.3 (b)

There are different ways of writing the equation of a line. The most usual of these is in the form $y = mx + c$. From this we can see directly the gradient of the line: it is the x-coefficient, m. If we substitute the value $x = 0$ into the equation, we find that $y = m \times 0 + c = c$. Since x is zero along the y-axis, we have found the point where the line crosses this axis: this point is called the y-intercept.

Figure 4.4

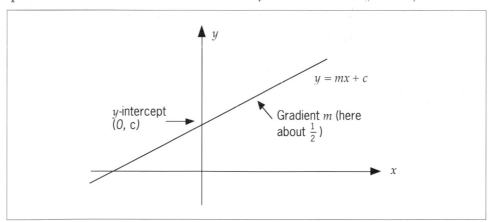

So in fact just from the equation we can make an immediate sketch of the line without any further calculation. Take the equation $y = 2x - 1$ for example. The x-coefficient is 2 which means that the gradient is 2, fairly steep. The y-intercept is –1, so this is where it crosses the y-axis.

Figure 4.5

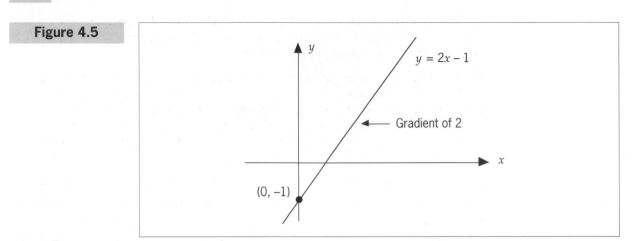

Similarly the equation $y = -\dfrac{1}{2}x + 3$ has a gradient of $-\dfrac{1}{2}$. This means that it slopes the other way, not very steeply. It crosses the y-axis where $y = 3$.

Figure 4.6

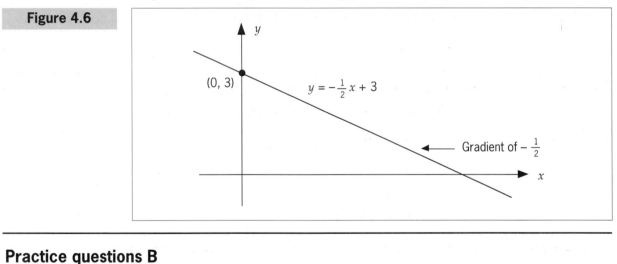

Practice questions B

1   Sketch the following lines, giving the coordinates of the point where the line crosses the y–axis:

(a) $y = 3x + 1$ (b) $y = -2x + 4$

(c) $y = x + 2$ (d) $y = -x + 1$

(e) $y = \dfrac{1}{3}x - 2$ (f) $y = -\dfrac{3}{2}x$

It might be useful at this point to look inside the head of a typical mathematician. As a breed, and this includes your future examiner, they like things to be neat and tidy. This has all kinds of implications in the way they like work to be presented, which we can look at as we go along, but for the moment it explains why there are different ways of writing an equation for the same straight line.

There is a definite hierarchy in their preference for different kinds of numbers – they prefer whole numbers to fractions, for example (but they prefer fractions to decimals, which is why they would write $\dfrac{2}{3}$ rather than 0.666...). Consequently, although they can accept an equation presented as $y = \dfrac{3}{4}x - \dfrac{1}{2}$, they are happier if the fractions are removed by multiplying throughout by 4 to give $4y = 3x - 2$.

In a similar fashion, they prefer expressions to begin with a positive term (so they would write $1 - x^2$ rather than $-x^2 + 1$). The equation $y = -x + 2$ for example, they might prefer to see as $x + y = 2$.

There is actually a different reason why an equation for a straight line can be written in the form $ax + by + c = 0$. This is because there is a corresponding equation in 3 dimensions (for a plane), which can be written in the form $ax + by + cz + d = 0$. In this way the relationship between equations in different dimensions can be seen as part of a general pattern rather than isolated parts having nothing in common with each other.

The result of all this is that what is basically the same equation can appear in different forms. For example, $y = -\frac{1}{2}x + 3$ is one form. Multiplying by 2 to clear the fraction gives $2y = -x + 6$ and bringing the x to the other side gives $x + 2y = 6$, which can also be written $x + 2y - 6 = 0$! You will generally find that the form required is specified in each particular question, and so you need to be able to change from one form to another.

Example	Clear the fractions in the following equations:

(a) $y = \frac{3}{4}x - 2$ (b) $2y = \frac{2}{3}x + \frac{1}{2}$

Solution	

(a)   Multiplying through by 4, $4y = 3x - 8$

(b)   This can be done either in two stages

$$\times 3 \implies 6y = 2x + \frac{3}{2}$$

$$\times 2 \implies 12y = 4x + 3$$

or in one go,

$$\times 6 \implies 12y = 4x + 3$$

Example	Express the following in the form $y = mx + c$, stating the gradient:

(a) $3x + 2y = 12$ (b) $4x - 5y + 17 = 0$

Solution	

(a)   Rearranging, $2y = -3x + 12$

$\div 2$ $y = -\frac{3}{2}x + 6$

Gradient is $-\frac{3}{2}$

(b)   Take y to the right hand side to make it positive

$$4x + 17 = 5y$$

$\div 5 \implies y = \frac{4}{5}x + \frac{17}{5}$

Gradient is $\frac{4}{5}$

Example	Express in the form $ax + by + c = 0$, where a, b and c are integers:

(a) $2y = 3x - 5$ (b) $2y = \frac{1}{3}x - \frac{3}{5}$

Solution

(a) Take y to the RHS so that the x stays positive

$$0 = 3x - 2y - 5 \quad \Rightarrow \quad 3x - 2y - 5 = 0$$

(b) $\times$ through by 15

$$30y = 5x - 9$$

$$\Rightarrow 5x - 30y - 9 = 0$$

Have a go at some of these.

Practice questions C

1 Multiply each of these equations by a suitable number to clear the fractions:

(a) $y = \dfrac{1}{2}x + 1$ (b) $y = \dfrac{2}{3}x - 2$ (c) $2y = \dfrac{4}{5}x - 3$

(d) $y = \dfrac{1}{3}x - \dfrac{1}{2}$ (e) $3y = \dfrac{1}{2}x - \dfrac{1}{4}$ (f) $2y = \dfrac{2}{5}x + \dfrac{1}{3}$

2 Express each of these equations in the form $y = mx + c$, stating the gradient of the line:

(a) $x + y = 6$ (b) $2y = 3x - 1$ (c) $2x + y + 5 = 0$ (d) $3x + y - 5 = 0$

(e) $3x + 2y = 6$ (f) $2x - y + 7 = 0$ (g) $-3y = x - 5$ (h) $4x - 3y + 12 = 0$

3 Express the following equations in the form $ax + by + c = 0$, where a, b and c are integers:

(a) $y = x - 3$ (b) $y = \dfrac{1}{2}x + 2$ (c) $2y = -3x + 7$

(d) $2y = \dfrac{1}{3}x - 1$ (e) $y = 3x - \dfrac{1}{2}$ (f) $3y = \dfrac{2}{5}x + \dfrac{4}{5}$

Finding the equation of a line   OCR P1 5.1.3 (b)

If we know the gradient of the line and a point through which it passes, we can find a relationship between the coordinates of any other point through which it passes. Suppose it has a gradient of 2 and passes through the point P with coordinates $(3, 1)$. If we take any other point Q on the line and give it the coordinates (x, y).

Figure 4.7

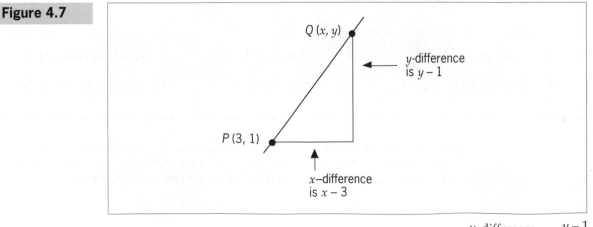

We can see that the gradient (using the definition gradient $= \dfrac{y\text{–difference}}{x\text{–difference}}$) is $\dfrac{y - 1}{x - 3}$

But we are told that the gradient of the line is 2, i.e. $\frac{y-1}{x-3} = 2$ and this is the relationship between the x- and y-coordinates of any other point on the line, in other words the equation of the line. We multiply through by $(x-3)$ to give

$$y - 1 = 2(x - 3)$$
$$\Rightarrow \quad y - 1 = 2x - 6$$
$$y = 2x - 5$$

This gives us a general method for finding the equation of a line given the gradient m and a point (x_1, y_1) through which the line passes. Taking a general point (x, y) as before, we have the y-difference $(y - y_1)$ and the x-difference $(x - x_1)$. This gives the gradient

$$m = \frac{y - y_1}{x - x_1}$$

Multiplying through by $x - x_1$, $y - y_1 = m(x - x_1)$. This is an important result:

> An equation of the straight line with gradient m,
>
> passing through the point (x_1, y_1) is
>
> $$y - y_1 = m(x - x_1)$$

Example

Find in the form $ax + by + c = 0$, where a, b and c are integers, the equation of the straight line with gradient $-\frac{1}{2}$, passing through $(2, 5)$.

Solution

Here $x_1 = 2$, $y_1 = 5$ and $m = -\frac{1}{2}$ so we have the equation

$$y - 5 = -\frac{1}{2}(x - 2)$$

$\times 2 \qquad 2y - 10 = -(x - 2) = -x + 2$

$\Rightarrow$ equation is $x + 2y - 12 = 0$

Practice questions D

1 Find an equation of the line with gradient:

(a) 2 passing through (1, 3) (b) $\frac{1}{2}$ passing through (2, –4) (c) –3 passing through (1, 1)

(d) $-\frac{1}{3}$ passing through (9, 5) (e) $\frac{3}{4}$ passing through (0, 0) (f) 0 passing through (2, 5)

We are frequently asked to find an equation of the line passing through a given pair of points. In this case, we find the gradient first of all and then use either of the given pair of points to substitute into the standard equation.

Example	Find the equation of the line passing through the points (2, 3) and (5, –3).

Solution	The gradient $m = \dfrac{-3-3}{5-2} = \dfrac{-6}{3} = -2$.

Taking the first point, the equation is

$$y - 3 = -2\,(x - 2) \implies y - 3 = -2x + 4$$

i.e. $2x + y - 7 = 0$

Practice questions E

1   Find an equation for the line passing through:

    (a)   (3, 2) and (5, 6)   (b)   (–3, 1) and (4, 8)   (c)   (0, 0) and (2, 4)

    (d)   (1, –1) and (–1, 1)   (e)   (2, 2) and (4, –4)   (f)   (2, 5) and (4, 5)

Parallel lines

OCR P1 5.1.3 (c)

Lines with the same gradient are called *parallel*. If we take three parallel lines, each with a gradient of 3, say $y = 3x + 1$, $y = 3x - 2$ and $y = 3x - 5$ we can see that they cross the y-axis at different points found by putting $x = 0$ into the equations (since $x = 0$ on the y-axis), or simply by looking at the constant term after the x-term in the equation.

Figure 4.8	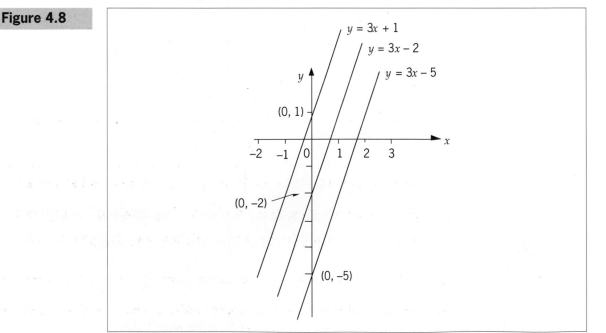

So if we have to find an equation of a line parallel to a given line, the only difference in the equation is in the constant term. We can simply rewrite the original equation with the unknown constant c in place of the previous constant and substitute the given point to find the value of this constant.

Example

Find the equation of the line parallel to:

(a) $y = 2x + 3$, passing through $(1, 1)$

(b) $2x + 5y + 7 = 0$, passing through $(2, 2)$

Solution

(a) The line parallel to $y = 2x + 3$ will have equation $y = 2x + c$.

Substituting $(1, 1)$ gives $1 = 2 \times 1 + c \Rightarrow c = -1$: equation is $y = 2x - 1$.

(b) Parallel line has equation $2x + 5y + c = 0$.

Substituting $(2, 2)$ $2 \times 2 + 5 \times 2 + c = 0$ $\Rightarrow$ $c = -14$

and the equation is $2x + 5y - 14 = 0$

Perpendicular lines

OCR P1 5.1.3 (c)

We have just seen that if two lines, $y = m_1 x + c_1$ and $y = m_2 x + c_2$, are parallel, they have the same gradient, i.e. $m_1 = m_2$.

Figure 4.9

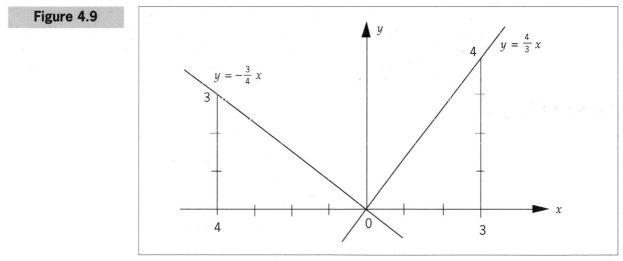

If we look at a part of the lines $y = \frac{4}{3}x$ and $y = -\frac{3}{4}x$ we can see that they are perpendicular to each other. While the line $y = \frac{4}{3}x$ goes along 3, it goes *up* 4. The line $y = -\frac{3}{4}x$ does just the opposite. As it goes along 4, it goes *down* 3.

The gradients of these lines are respectively $\frac{4}{3}$ and $-\frac{3}{4} = -\dfrac{1}{\frac{4}{3}}$. This is true for any two perpendicular lines: if the gradient of one is m_1, then the gradient of the other, m_2 say, is $\dfrac{-1}{m_1}$ i.e. $m_2 = \dfrac{-1}{m_1} \Rightarrow m_1 m_2 = -1$.

This is a very important result and is summarised as follows.

> If two lines have gradients m_1 and m_2 then
>
> (a) $m_1 = m_2$ $\Leftrightarrow$ lines are parallel
>
> (b) $m_1 m_2 = -1$ $\Leftrightarrow$ lines are perpendicular

Example

(a) Show that the lines $y = 4x - 3$ and $x + 4y - 8 = 0$ are perpendicular.

(b) Find an equation of the line perpendicular to the line $3x + 4y - 5 = 0$, passing through the point $(1, 1)$.

Solutions

(a) The first equation is in the form $y = mx + c$ and we can read off the gradient m as 4.

After rearranging, the second equation is $y = -\frac{1}{4}x + 2$ and so the gradient is $-\frac{1}{4}$.

$4 \times -\frac{1}{4} = -1$, so the lines are perpendicular.

(b) Rearranging to $y = -\frac{3}{4}x + \frac{5}{4}$, we can see that the gradient of $3x + 4y - 5 = 0$ is $-\frac{3}{4}$, so we want the gradient of the other line, m say, to be such that

$$\left(-\frac{3}{4}\right) \times m = -1 \quad \Rightarrow \quad m = \frac{4}{3}$$

The required equation is $y - 1 = \frac{4}{3}(x - 1) \Rightarrow \quad 3y - 3 \quad = 4x - 4$

$$3y \quad = 4x - 1$$

Example

Find the equation of the line joining $A(-1, -9)$ to $B(6, 12)$.

Another line passes through $C(7, -5)$ and meets AB at right angles at D.

Find the equation of CD.

Solution

The gradient of the line joining these points is:

$$\frac{12 - (-9)}{6 - (-1)} = \frac{21}{7} = 3.$$

and we can take either point as (x_1, y_1) say $A(-1, -9)$.

Then the required equation is:

$$y - (-9) = 3(x - (-1)) \Rightarrow y + 9 = 3x + 3 \quad \text{i.e. } y = 3x - 6$$

Figure 4.10

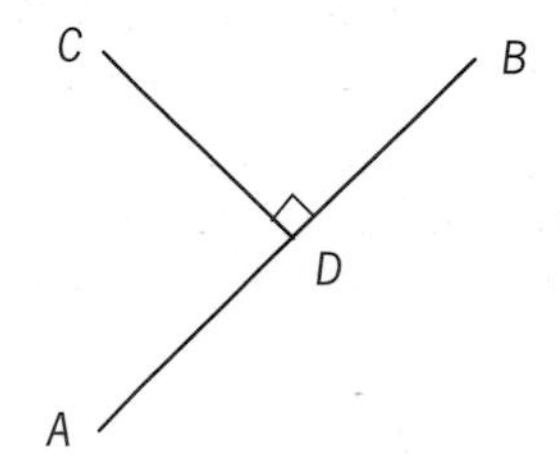

The gradient of AB is 3, so the equation of the line CD, perpendicular to AB, is $-\frac{1}{3}$.

Passing through $C(7, -5)$, it has equation, using the standard formula,

$$y - (-5) = -\frac{1}{3}(x - 7) \Rightarrow y + 5 = -\frac{1}{3}x + \frac{7}{3}$$

i.e. $3y + 15 = -x + 7 \Rightarrow x + 3y + 8 = 0$

Practice questions F

1 State whether the following pairs of lines are parallel, perpendicular, or neither:

(a) $y = 2x + 3$, $y = 2x$

(b) $x + y = 3$, $x + 2y = 5$

(c) $y = 3x - 1$, $x + 3y = 7$

(d) $y = 5x + 7$, $5x - y + 11 = 0$

(e) $y = 2x - 5$, $y = \frac{1}{2}x - 1$

(f) $3x + 4y - 12 = 0$, $3y = 4x - 1$

(g) $2y = x + 3$, $y = 5 - 2x$

(h) $x + 2y - 5 = 0$, $4x - 2y + 7 = 0$

2 Find an equation for the line parallel to:

(a) $y = 2x + 9$ passing through $(2, 2)$

(b) $y = -3x + 5$ passing through $(3, -2)$

(c) $x + y = 4$ passing through $(3, -2)$

(d) $3x + 2y - 6 = 0$ passing through $(2, -3)$

(e) $x + 5y + 7 = 0$, passing through $(-2, 1)$

3 Find an equation for the line perpendicular to:

(a) $y = 3x - 1$ passing through $(6, 1)$

(b) $y = -\frac{1}{2}x + 2$ passing through $(1, 1)$

(c) $y = 2 - 4x$ passing through $(8, 3)$

(d) $x + y - 7 = 0$ passing through $(2, -1)$

(e) $2x + 3y + 2 = 0$ passing through $(4, 7)$

(f) $3x - 4y + 1 = 0$ passing through $(9, 4)$

Mid-points

OCR P1 5.1.3 (a)

To find the mid-point of the line joining two points, we take the *average* of the x-coordinates and the average of the y-coordinates. For example, the mid-point of the line joining the points $(-1, 4)$ and $(7, -2)$ would be

$$\frac{-1 + 7}{2}, \frac{4 + (-2)}{2} \quad \text{i.e.} \quad (3, 1).$$

(A common mistake here, by the way is to *subtract* the x and y-coordinates.)

Example	For the points $A(2, 3)$ and $B(0, 7)$, find:

(a) the mid-point M (b) the gradient of the line.

Deduce the gradient of a line perpendicular to AB and hence find an equation for the line passing through M and perpendicular to AB. (This line is called the *perpendicular bisector* of the line AB.)

Solution	It quite often helps with this type of equation to draw a rough sketch of the situation.

Figure 4.11

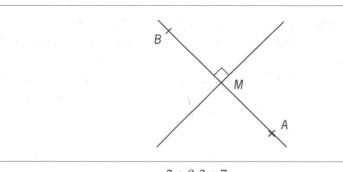

(a)   The mid-point M of AB is $\dfrac{2+0}{2}, \dfrac{3+7}{2}$ i.e. $(1, 5)$.

(b)   The gradient of AB is $\dfrac{7-3}{0-2} = -2$ which means that a line perpendicular to AB has gradient $\dfrac{1}{-2} = -\dfrac{1}{2}$. In particular, the line through $M(1, 5)$ has equation

$$y - 5 = \dfrac{-1}{2}\,(x - 1)$$

$$2y - 10 = -x + 1$$

$$\Rightarrow \qquad x + 2y - 11 = 0$$

Practice questions G

1   Find the mid-point of the line joining the points:

(a) $(1, 2)$ and $(3, 6)$ (b) $(0, -1)$ and $(4, 3)$ (c) $(-2, 5)$ and $(7, 9)$

(d) $(-1, 2)$ and $(2, -1)$ (e) $(3, 5)$ and $(-3, -1)$ (f) $(2, -2)$ and $(-3, -4)$

2   Find an equation of the perpendicular bisector of the lines joining the points in Question 1.

SUMMARY EXERCISE

1   The points A, B and C have coordinates $(5, -3)$, $(7, 8)$ and $(-3, 4)$ respectively. The midpoint of BC is M.

(a)   Write down the coordinates of M.

(b)   Find the equation of the straight line which passes through the points A and M.

[AEB 1996]

2   The straight lines with cartesian equations $3x + 2y = 1$ and $2x + 5y = 19$ intersect at the point P.

(a)   Calculate the coordinates of P.

(b)   Determine an equation for the line through the point Q $(7, 3)$ which is perpendicular to the line with equation $2x + 5y = 19$.

[AEB 1997]

3   The line L passes through the points A $(1, 3)$ and B $(-19, -19)$.

(a)   Calculate the distance between A and B.

(b)   Find an equation of L in the form $ax + by + c = 0$, where a, b and c are integers.

4   (a)   Find an equation of the straight line passing through the points with coordinates $(-1, 5)$ and $(4, -2)$, giving your answer in the form $ax + by + c = 0$, where a, b and c are integers.

The line crosses the x-axis at the point A and the y-axis at the point B, and O is the origin.

(b)   Find the area of $\triangle\,OAB$.

5   The points A and B have coordinates (2, 16) and (12, –4) respectively. A straight line l_1 passes through A and B.

   (a)   Find an equation for l_1 in the form
   $ax + by = c$.
   The line l_2 passes through the point C with coordinates (–1, 1) and has gradient $\frac{1}{3}$.

   (b)   Find an equation for l_2.
   The lines l_1 and l_2 intersect at the point D. The point O is the origin.

   (c)   Find the length OD, giving your answer in the form $m\sqrt{5}$, where m is an integer.

6   The straight line l with equation $12x - 5y + 10 = 0$ cuts the x-axis at A and the y-axis at B. The origin is at O.

   (a)   Calculate the area of the triangle OAB.

   (b)   Find the equation of the straight line which passes through O and which is perpendicular to the line l.

   (c)   Find the length of AB and hence, or otherwise, calculate the shortest distance from O to the line l.   [AQA 1999]

7   (a)   Find an equation of the line l which passes through the points $A(1, 0)$ and $B(5, 6)$.
   The line m with equation $2x + 3y = 15$ meets l at the point C.

   (b)   Determine the coordinates of C.
   The point P lies on m and has x-coordinate –3.

   (c)   Show, by calculation, that $PA = PB$.

8   The straight line l_1 passes through the points A and B with coordinates (2, 2) and (6, 0) respectively.

   (a)   Find an equation of l_1.
   The straight line l_2 passes through the point C with coordinates (–9, 0) and has gradient $\frac{1}{4}$.

   (b)   Find an equation of l_2.
   The lines l_1 and l_2 intersect at the point D.

   (c)   Calculate, to 2 decimal places, the length of AD.

   (d)   Calculate the area of $\triangle DCB$.

9   The straight line l passes through A (1, 3√3) and B (2 + √3, 3 + 4√3).

   (a)   Calculate the gradient of l giving your answer as a surd in its simplest form.

   (b)   Give the equation of l in form $y = mx + c$, where constants m and c are surds to be given in their simplest form

   (c)   Show that l meets the x-axis at the point C (–2, 0).

   (d)   Calculate the length of AC.

   (e)   Find the size of the acute angle between the line AC and the x-axis, giving your answer in degrees.

10 $ABCD$ is a parallelogram whose diagonals meet at M. The coordinates of A and B are (1, 5) and (4, 2) respectively. Given that BD is parallel to $5x + y = 0$ and that AC is perpendicular to $5x - y = 0$, find the equation of BD and of AC. Calculate:

   (a)   the coordinates of M

   (b)   the coordinates of C and of D.
   Prove that $ABCD$ is a rectangle.

SUMMARY   When you have finished this section you should:

- know the meaning of the gradient of a line
- know that the standard equation of a line is $y = mx + c$ and understand the significance of the constants m and c
- be familiar with other forms of the equation and be able to change from one to another
- know how to find the gradient of a line joining two given points

- know how to find the equation of a line passing through:
  - (a) a given point with a given gradient
  - (b) two given points
- know the condition for two lines to be
  - (a) parallel
  - (b) perpendicular
- know how to find the midpoint of the line joining two given points
- be able to find the perpendicular bisector of the line joining two given points
- be able to find the distance between two points
- be able to find the point where two lines intersect.

ANSWERS

Practice questions A

1. (a) 2   (b) 1   (c) 2   (d) −1
   (e) −3   (f) $\frac{1}{2}$ (g) −8   (h) $\frac{-7}{3}$
   (i) 0   (j) ∞

Practice questions B

1. (a)

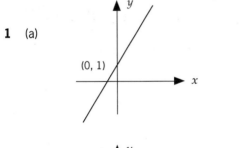

(b)

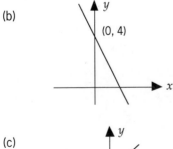

(c)

(d)

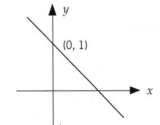

(e)

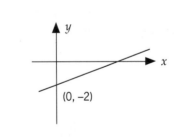

(f)

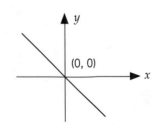

Practice questions C

1 (a) $2y = x + 2$

(b) $3y = 2x - 6$

(c) $10y = 4x - 15$

(d) $6y = 2x - 3$

(e) $12y = 2x - 1$

(f) $30y = 6x + 5$

2 (a) $y = -x + 6, -1$

(b) $y = \frac{3}{2}x - \frac{1}{2}, \frac{3}{2}$

(c) $y = 2x - 5, -2$

(d) $y = -3x + 5, -3$

(e) $y = -\frac{3}{2}x + 3, -\frac{3}{2}$

(f) $y = 2x + 7, 2$

(g) $y = -\frac{1}{3}x + \frac{5}{3}, -\frac{1}{3}$

(h) $y = \frac{4}{3}x + 4, \frac{4}{3}$

3 (a) $x - y - 3 = 0$

(b) $x - 2y + 4 = 0$

(c) $3x + 2y - 7 = 0$

(d) $x - 6y - 3 = 0$

(e) $6x - 2y - 1 = 0$

(f) $2x - 15y + 4 = 0$

Practice questions D

1 (a) $y = 2x + 1$

(b) $y = \frac{1}{2}x - 5$

(c) $y = -3x + 4$

(d) $y = -\frac{1}{3}x + 8$

(e) $y = \frac{3}{4}x$

(f) $y = 5$

Practice questions E

1 (a) $y = 2x - 4$

(b) $y = x + 4$

(c) $y = 2x$

(d) $y = -x$

(e) $y = -3x + 8$

(f) $y = 5$

Practice questions F

1 (a) parallel

(b) neither

(c) perpendicular

(d) parallel

(e) neither

(f) perpendicular

(g) perpendicular

(h) perpendicular

2 (a) $y = 2x - 2$

(b) $y = -3x + 4$

(c) $x + y = 1$

(d) $3x + 2y = 0$

(e) $x + 5y - 3 = 0$

3 (a) $x + 3y = 9$

(b) $y = 2x - 1$

(c) $4y = x + 4$

(d) $y = x - 3$

(e) $2y = 3x + 2$

(f) $4x + 3y = 48$

Practice questions G

1 (a) $(2, 4)$

(b) $(2, 1)$

(c) $(\frac{5}{2}, 7)$

(d) $(\frac{1}{2}, \frac{1}{2})$

(e) $(0, -2)$

(f) $(-\frac{1}{2}, -3)$

2 (a) $y = -\frac{1}{2}x + 5$

(b) $x + y = 3$

(c) $y = \frac{-9}{4}x + \frac{101}{8}$

(d) $y = x$

(e) $x + y + 3 = 0$

(f) $y = \frac{-5}{2}x - \frac{17}{4}$

SECTION 5

Sequences and series

INTRODUCTION
When we study mathematics we frequently find ourselves searching for a pattern which relates different items of a collection to each other. In this section we shall be concerned with sequences and series of numbers – finding the pattern behind them and using this to find concise ways of describing them. There will be an introduction to the symbol Σ for the sum of a series, a useful device which you will meet again further on in the course. We shall then look at two standard types of series – arithmetic and geometric, devising and using general formulae by which we can find terms and sums, and seeing how these can be applied to practical situations.

Sequences

OCR P2 5.2.2 (a)

If we look at the sequences of numbers:

2, 4, 6, 8 …

we can see that there are two different ways of describing it. Either we notice that each number is two more than the previous number or we notice that the value of any particular term is twice the value of its position, so that the 13th term for example is $2 \times 13 = 26$.

It's useful when we're trying to investigate sequences to have some standard way of referring to the terms, so conventionally we write the first term as u_1, the second as u_2, the third as u_3 and so on. In the sequence above,

$$u_1 = 2 \quad u_2 = 4 \quad u_3 = 6 \quad u_4 = 8$$

In general, we write the nth term as u_n. Then the two ways of describing the sequence are:

(a) The nth term is twice the previous term, the $(n-1)^{\text{th}}$ term,

i.e. $u_n = 2 \times u_{n-1}$

(b) the nth term is twice the position value,

i.e. $u_n = 2n$

For the moment we are only going to look at sequences which are described by the second of these methods, i.e. the nth term is given in terms of a formula involving n.

Example

Write out the first four terms of the sequences defined by:

(a) $u_n = 3n - 1$ (b) $u_n = \dfrac{n}{n+1}$ (c) $u_n = 2^{n+1}$

Solution	(a) To find u, we substitute $n = 1$ into the defining formula. This gives

$$u_1 = 3(1) - 1 = 2$$

Similarly
$$u_2 = 3(2) - 1 = 5$$
$$u_3 = 3(3) - 1 = 8$$
$$u_4 = 3(4) - 1 = 11$$

so the sequence is 2, 5, 8, 11, ...

(b) $u_1 = \dfrac{1}{1+1} = \dfrac{1}{2}$; $u_2 = \dfrac{2}{2+1} = \dfrac{2}{3}$; $u_3 = \dfrac{3}{3+1} = \dfrac{3}{4}$; $u_4 = \dfrac{4}{4+1} = \dfrac{4}{5}$

and the sequence is $\dfrac{1}{2}, \dfrac{2}{3}, \dfrac{3}{4}, \dfrac{4}{5}, \ldots$

(c) $u_1 = 2^{1+1} = 2^2 = 4$: $u_2 = 2^3 = 8$: $u_3 = 2^4 = 16$: $u_4 = 2^5 = 32$

i.e.   4, 8, 16, 32, ...

Practice questions A

1   Write out the first four terms of the sequence defined by

(a) $u_n = 2n + 5$

(b) $u_n = \dfrac{1}{n+3}$

(c) $u_n = 3^n$

(d) $u_n = \dfrac{n+1}{n+3}$

(e) $u_n = n(n-1)$

(f) $u_n = n^2 + 5n$

(g) $u_n = n \times 2^n$

(h) $u_n = \dfrac{2^{n+1}}{3^{n-1}}$

Finding the general term

Sometimes we have to work the other way round: we are given the sequence and asked to find the formula for the nth term (or sometimes r^{th} term or i^{th} term). The first thing to have a look at is the difference between pairs of successive terms: if this is constant the formula can be found without any trouble since it has to involve the term $n \times$ difference

Example	Find the nth term of the sequence

3, 7, 11, 15, ...

Solution	The difference between the pairs of terms is 4, so the formula involves the term $4n$. We then go back one term before the beginning, i.e. we have to subtract 4 from the first term to give $3 - 4 = -1$. This is what must be added to the term involving n to give the nth term,

i.e. $u_n = 4n - 1$

You can check that this does in fact give us the correct sequence.

Example	Find the rth term of the sequence

15, 12, 9, 6, ...

| Solution | Here the difference is –3, so we have –3r. Going back one, $15 + 3 = 18$ and so the rth term is |

$$u_r = 18 - 3r$$

When the difference is not constant things are not so simple, and we have to rely on recognising some standard sequence.

| Example | Find the nth term of the sequence |

$$2, 5, 10, 17, \ldots$$

| Solution | The difference is no longer constant but we might be able to see that we could rewrite the sequence as |

$$1 + 1, \ 4 + 1, \ 9 + 1, \ 16 + 1, \ \ldots$$

and recognise the first in each pair as a square number, i.e.

$$1^2 + 1, \ 2^2 + 1, \ 3^2 + 1, \ 4^2 + 1, \ \ldots$$

and so $\quad u_n = n^2 + 1$

Practice questions B

1 Find an expression for the nth term of the following sequences

(a) 4, 7, 10, 13, …

(b) 10, 6, 2, –2, …

(c) $\frac{1}{2}, \frac{2}{3}, \frac{3}{4}, \frac{4}{5}, \ldots$

(d) $1, 1\frac{1}{2}, 2, 2\frac{1}{2}, \ldots$

(e) 0, 3, 8, 15, …

(f) 1, 8, 27, 64, …

Series

OCR P2 5.2.2 (b)

Expressions such as:

$$1 + 3 + 5 + 7 + 9 \quad \text{and} \quad 4 + 2 + 1 + \frac{1}{2} + \frac{1}{4} + \ldots$$

are called *series*: they are sums of sequences. There are different types of series, depending on the relationship between successive pairs of terms.

In the first series, you can see that a fixed number, in this case 2, has been *added* to each term to get the next term. Series like these are *arithmetic*. When the terms are *multiplied* by a certain number to get the next term, as in the second series where each term has been multiplied by $\frac{1}{2}$, the series is *geometric*. We shall be looking at these more closely, but first of all we shall introduce a piece of mathematical shorthand, the Σ-notation.

Σ notation

Σ, pronounced 'sigma', is the Greek letter S, used to stand for *sum*. In order to write down the first series we looked at, i.e. $1 + 3 + 5 + 7 + 9$ with $u_1 = 1$, $u_2 = 3$ etc, we need the *general* term, in this case $u_r = 2r - 1$. Then we decide where we want to start, $r = 1$ and finish, $r = 5$, and write:

$$\sum_{r=1}^{5} (2r - 1)$$

In English this means 'put the first value of r into the general term, then the next, then the next and keep going until you reach the top limit. Then add all these terms together'.

Other examples would be:

$$\sum_{r=2}^{10} (r^2 + 1) = \underset{r=2}{5} + \underset{r=3}{10} + \underset{r=4}{17} + \ldots + \underset{r=10}{101}$$

and $$\sum_{r=0}^{8} \frac{1}{2^r} = \underset{r=0}{\frac{1}{2^0}} + \underset{r=1}{\frac{1}{2}} + \underset{r=2}{\frac{1}{2^2}} + \ldots + \underset{r=8}{\frac{1}{2^8}}$$

We have to be careful if we want to find the number of terms – there's a tendency with something like $\sum_{r=2}^{10} (r^2 + 1)$ to say that the number of terms is simply the top limit minus the bottom limit, i.e. $10 - 2 = 8$. Actually, because both the top and bottom terms are included, there are in fact $8 + 1 = 9$ terms. Another example would be $\sum_{r=n}^{2n} r^2$: here the number of terms is $2n - n + 1 = n + 1$, and not n, as you might think.

Practice questions C

1 Write down the first three and last two terms of the following series

(a) $\sum_{r=1}^{10} (2r + 1)$ (b) $\sum_{r=0}^{7} 3^r$

(c) $\sum_{i=5}^{12} i(i + 1)$ (d) $\sum_{i=1}^{20} \frac{i + 3}{i + 1}$

(e) $\sum_{n=0}^{10} (n^2 + n)$ (f) $\sum_{n=2}^{8} n \times 2^n$

2 Find the number of terms in the following sequences

(a) $\sum_{r=1}^{7} r^2$ (b) $\sum_{i=5}^{10} i(i + 4)$

(c) $\sum_{r=10}^{30} \frac{r}{r + 1}$ (d) $\sum_{r=n}^{2n} (r + 3)$

Arithmetic progressions

OCR P2 5.2.2 (b)

Let's write down some examples of the first of these two types of series, arithmetic progressions (APs for short). If we start with any number, say 5, and add a second number, say 3, and add the 3 again, and again, and again… we'd end up with a sequence of terms looking like this:

5, 8, 11, 14, 17, 20, …

The first term doesn't have to be a whole number and the number that we're adding doesn't have to be positive. We could start, for example, with $3\frac{1}{4}$ and add -1 each time and then this would give:

$3\frac{1}{4}, 2\frac{1}{4}, 1\frac{1}{4}, \frac{1}{4}, -\frac{3}{4}, -1\frac{3}{4}, \ldots$

Here are some more examples of this type of progression:

(a)   1, 3, 5, 7, 9, …

(b)   3, 3.3, 3.6, 3.9, 4.2, …

(c) $p,\ p + q,\ p + 2q,\ p + 3q,\ p + 4q,\ …$

The fixed amount that is added to each term to give the next is called the *common difference* and is written d. When d is positive, the sequence is increasing; when it's negative, the terms decrease. The first term is usually written a, so any arithmetic progression can be written:

$$a,\ a + d,\ a + 2d,\ a + 3d,\ a + 4d,\ …$$

with different values of a and d. (So that the series 5, 8, 11, 14, … which we put together first of all has $a = 5$ and $d = 3$, the next has $a = 3\frac{1}{4}$ and $d = -1$.) Now the second term is $a + d$ and the fourth is $a + 3d$, so we can see that the number of ds lags one behind the number of the term. If we write the nth term as u_n, then:

$$u_n = a + (n - 1)\, d$$

Summing the series
OCR P2 5.2.2 (c)

[The following proof is important for Edexcel candidates, as you are required to know it as part of the Edexcel P1 syllabus – make sure you understand it and can reproduce it.]

We'll now see what happens when we add the first n of these terms together. Writing this sum as S_n, we get:

$$S_n = a + (a + d) + (a + 2d) + (a + 3d) + … + [a + (n - 1)d] \quad [1]$$

If we write this sum the other way round, we have:

$$S_n = [a + (n - 1)d]\ + [a + (n - 2)d] + … + [a + d] + a \quad\quad [2]$$

We now add these two series by adding the first term of the first series with the first term of the second series, i.e. $a + [a (n - 1)d] = 2a + (n - 1)d$. Then the second terms of each, i.e. $[a + d] + [a + (n - 2)d] = 2a + (n - 1)d$, and continue in this way until we add the last term of each series, i.e. $[a + (n - 1)d] + a = 2a + (n - 1)d$. Each pair of terms has the same sum, i.e. $2a + (n - 1)d$ and since there are n pairs, they total $n[2a + (n - 1)d]$. We have added S_n to S_n so

$$2S_n = n[2a + (n - 1)d]$$

$$S_n = \frac{n}{2}[2a + (n - 1)d] \quad\quad\quad\quad\quad\quad [3]$$

This is one way of expressing the sum to n terms of an arithmetic series. If we look at the first and last terms of the series [1], a and $a + (n - 1)d$ respectively, we can see that if we added them together we would have the contents of the brackets in [3], i.e.

$$S_n = \frac{n}{2}\, \{\text{first term} + \text{last term}\}$$

and this can be a useful alternative way of expressing the sum. The box below contains the general properties we've found so far:

> If u_n is the nth term of an AP,
>
> $$u_n = a + (n-1)d$$
>
> If S_n is the sum of the first n terms of an AP,
>
> $$S_n = \frac{n}{2}\{2a + (n-1)d\}$$
>
> $$= \frac{n}{2}\{\text{first term + last term}\}$$

Let's have a look at some examples of this:

Example	Write down the first five terms of the AP with $a = 3$ and $d = 2$. Find the tenth term and the sum of the first 20 terms.

Solution	The sequence will be:

$$3, 5, 7, 9, 11, \ldots$$

$$\begin{aligned}
\text{The tenth term, } u_{10} &= a + (10-1)d \\
&= 3 + 9 \times 2 \quad = 21
\end{aligned}$$

For the sum, using $S = \frac{n}{2}\{2a + (n-1)\,d\}$ with $n = 20$, $a = 3$ and $d = 2$

$$\begin{aligned}
S_{20} &= \frac{20}{2}\{2 \times 3 + 19 \times 2\} \\
&= 10\,\{6 + 38\} \ = 440
\end{aligned}$$

Example	Find the twentieth term and the sum of the first 30 terms of the progression beginning with

$$10, 8, 6, 4, 2, \ldots$$

Solution	Here $\quad a = 10$ and $d = -2$

The twentieth term, $u_{20} = a + 19d$

$$= 10 + 19\,(-2) \ = -28$$

The sum of the first 30 terms, $S_{30} = \frac{30}{2}\{2a + 29d\}$

$$= 15\,\{20 + (-58)\} \ = -570$$

Example	Find the sum of the first 50 even integers.

Solution	The fiftieth even integer is 100 – the first is, of course, 2.

Using $\quad S_n = \frac{n}{2}\{\text{first term + last term}\}$

$$S_{50} = \frac{50}{2}\{2 + 100\} \quad = 25\,(102) \quad = 2550$$

Instead of being asked to find the sum of a series which we are told is arithmetic, the question could be put in Σ-notation, as in the next example.

Example	Find $\displaystyle\sum_{r=1}^{10} (2r + 3)$

| **Solution** | The first four terms are 5, 7, 9, 11, ..., i.e. an AP with $a = 5$ and $d = 2$ |

$$S_{10} = \frac{10}{2} \{2 \times 5 + (10 - 1)\, 2\} = 140$$

Practice questions D

In these questions, AP stands for arithmetic progression, u_n is the nth term and S_n is the sum of the first n terms.

1   In an AP, $u_2 = 15$ and $u_4 = 23$. Find S_{10}.

2   Find the sum of the even numbers between 20 and 80 inclusive.

3   (a)   Find the sum of the numbers between 1 and 100 which are multiples of 3.
    (b)   Find the sum of the numbers between 1 and 100 which are *not* multiples of 3

4   In the AP, 1, 7, 13, ... , $u_n = 49$. Find n.

5   In an AP, $u_4 = 18$ and $u_7 = 30$. Find u_1

6   In an AP, $u_n = 6n + 5$. Find u_3.

7   Find the value of the first term of the AP 100, 94, 88, ... to be negative.

8   In an AP, $u_n = 4n - 7$. Find S_{40}.

9   In an AP, $u_4 = 9$ and $S_4 = 21$. Calculate S_8.

10   In an AP, $S_8 = 56$ and $S_{20} = 260$. Calculate u_1 and the common difference.

11   A circle is completely divided into n sectors in such a way that the angles of the sectors are in AP. If the smallest of these angles is 8° and the largest 52°, calculate n.

12   In an AP, $u_3 = -20$ and $u_{11} = 20$. Find the common difference and first term.

When the first k terms of this series are added together, their sum is zero. Find the non-zero value of k.

13   In an AP, the sum of the first 10 terms is the same as the sum of the next 5 terms. Given that the first term is 12, find the sum of the first 15 terms.

14   A length of 200 cm is divided into 25 sections whose lengths are in AP. Given that the sum of the lengths of the 3 smallest sections is 4.2 cm, find the length of the largest section.

15   In an AP, $u_7 = 3u_2$ and $u_{10} = 21$. Find u_1.

16   In an AP, $u_1 = 3$ and the common difference is 0.8. If $S_n = 231$, find the value of n.

17   The angles of a triangle are in AP. If the common difference is 30°, find the angles.

18   Find the sum of the integers between m and n inclusive.

Deducing the *n*ᵗʰ term

The sum of the first four terms of a series, S_4, is $u_1 + u_2 + u_3 + u_4$ and the sum of the first three terms S_3 is $u_1 + u_2 + u_3$. If we subtract these expressions, we have:

$$S_4 - S_3 = (u_1 + u_2 + u_3 + u_4) - (u_1 + u_2 + u_3) = u_4$$

This is true in general: the difference between the sums of the first n terms and the sum of the first $(n - 1)$ terms is just the final term, u_n

$$u_n = S_n - S_{n-1}$$

Example:	The sum of the first n terms of a progression is given by

$$S_n = n(2n+1)$$

(a) Find an expression for S_{n-1} and deduce an expression for u_n

(b) Find the first term and the tenth term

Solution	(a) We replace n in the expression for S_n by $(n-1)$ which gives

$$
\begin{aligned}
S_{n-1} &= (n-1)\left[2(n-1)+1\right] \\
&= (n-1)(2n-1) \\
u_n &= S_n - S_{n-1} \\
&= n(2n+1) - (n-1)(2n-1) \\
&= 2n^2 + n - \left[2n^2 - 3n + 1\right] \\
&= 4n-1
\end{aligned}
$$

(b) The 1st term, $u_1 = 4 \times 1 - 1 = 3$

and the 10th term, $u_{10} = 4 \times 10 - 1 = 39$

Practice questions E

1 The sum S_n of the first n terms of a certain arithmetic progression is given by

$$S_n = n^2 + 3n$$

Find the first term and the common difference.

2 The formula

$$S_n = \left[\frac{n}{2}(n+1)\right]^2$$

gives the sum, S_n, of the first n terms of a certain series. Calculate the fifth term of the series.

Geometric progressions

OCR P2 5.2.2 (b),(c)

Remember that in a geometric progression (GP) we multiply any term by a fixed number (called the common ratio and written r) to give the next term. Some examples of this would be:

(a) 1, 2, 4, 8, 16, …

(b) 12, 3, $\frac{3}{4}$, $\frac{3}{16}$, $\frac{3}{64}$ …

(c) 5, −10, 20, −40, 80, …

(d) a, a^3, a^5, a^7, a^9, …

(e) 1, 1.1, 1.21, 1.331, 1.4641, …

From these you can see that r can be positive or negative, greater or less than 1. With the first term still written as a, sequence (a) has $a = 1$ and $r = 2$, and sequence (b) has $a = 12$ and $r = \frac{1}{4}$. You can probably work out a and r for the remaining three sequences.

The general progression can be written:

$$a, ar, ar^2, ar^3, ar^4, \ldots$$

and this time it's the power of r that lags one behind the number of the term (so that the fifth term is ar^4). The nth term is going to be:

$$u_n = ar^{n-1}$$

Write out the sum of the first n terms:

$$S_n = a + ar + ar^2 + ar^3 + \dots + ar^{n-1}$$

$$= a(1 + r + r^2 + r^3 + \dots + r^{n-1})$$

Multiply both sides of this equation by $(1 - r)$:

$$S_n(1-r) = a(1 + r + r^2 + r^3 + \dots + r^{n-1})(1-r)$$

$$= a[1 + r + r^2 + r^3 + \dots + r^{n-1} - r - r^2 - r^3 - \dots r^{n-1} - r^n]$$

$$= a(1 - r^n)$$

Then:
$$S_n = \frac{a(1 - r^n)}{1 - r}$$

This time there's no alternative way of writing the sum, although if $r > 1$ we can rewrite this as:

$$S_n = \frac{a(r^n - 1)}{r - 1} \text{, so that the denominator is positive.}$$

As with the proof for the sum of n terms of an AP, you have to know this derivation for the P1 syllabus.

Exercise	In a GP, $u_1 = 2\frac{1}{2}$ and $u_2 = 5$. Find S_5

Solution	

$$u = a = 2\tfrac{1}{2} \qquad \text{①}$$
$$u_2 = ar = 5 \qquad \text{②}$$

② ÷ ① gives $r = 2$

$$S_5 = \frac{a(r^5 - 1)}{r - 1} = \frac{2\frac{1}{2}(2^5 - 1)}{2 - 1} = 77\tfrac{1}{2}$$

Exercise	In a GP with positive terms, $u_5 = 45$ and $u_7 = 5$. Find u_6 and u_1.

Solution	

$$u_5 = ar^4 = 45 \qquad \text{①}$$
$$u_7 = ar^6 = 5 \qquad \text{②}$$

② ÷ ① $\dfrac{ar^6}{ar^4} = \dfrac{5}{45} \Rightarrow r^2 = \dfrac{1}{9} \Rightarrow r = \pm\dfrac{1}{3}$

But terms are positive, so $r = \dfrac{1}{3}$

$$u_6 = ar^5 \text{ but also } u_6 = u_5 \times r$$
$$= 45 \times \tfrac{1}{3} = 15$$

From ①, $ar^4 = 45 \Rightarrow a \times \left(\tfrac{1}{3}\right)^4 = 45$

$$\Rightarrow a = 45 \times 3^4 = 3645$$

Example	The lengths of the sides of a triangle are in GP. The length of the shortest side is 6 cm and the perimeter of the triangle is $28\frac{1}{2}$ cm. Find the lengths of the other sides.

| **Solution** | The length of the three sides are in GP so we can say that they are a, ar and ar^2. |

Then we have $a = 6$ ①

and $a + ar + ar^2 = 28\frac{1}{2}$

i.e. $a(1 + r + r^2) = 28\frac{1}{2}$ ②

Substitute ① into ② to give

$$6(1 + r + r^2) = 28\frac{1}{2}$$

$$\Rightarrow 1 + r + r^2 = \frac{19}{4} \quad \times \text{ by } 4$$

$$4 + 4r + 4r^2 = 19$$

$$\Rightarrow 4r^2 + 4r - 15 = 0$$

$$(2r - 3)(2r + 5) = 0$$

$$\Rightarrow r = \frac{3}{2} \text{ or } r = -\frac{5}{2}$$

Since the sides have positive lengths, $r = \frac{3}{2}$ and the sides are 6, 9 and $13\frac{1}{2}$ cm.

Practice questions F

In the following questions, GP stands for Geometric progression, u_n is the nth term and S_n is the sum of the first n terms.

1   Find u_{10} and S_{10} for the GP 1, 2, 4, 8, …

2   In a GP, $u_3 = 18$ and $u_5 = 162$. Find u_1.

3   In a GP with positive terms, $u_2 = 17\frac{1}{2}$ and $u_3 = 4\frac{2}{3}$. Calculate the value of the common ratio.

4   Find the number x so that 3, x, 4 are in GP.

5   In a GP, $u_3 = 2$ and $u_5 = \frac{1}{2}$. Find the possible values for the common ratio and find the corresponding values for u_6.

6   In a GP, $u_5 = 5$ and $u_6 = \frac{5}{2}$. Find S_{10} to one decimal place.

7   In a GP, $u_n = \left(\frac{2}{3}\right)^{n-1}$. Find u_1, u_3 and S_5.

8   Find $\sum_{r=1}^{10} 2 \times \left(\frac{3}{4}\right)^{r-1}$ to three decimal places.

Sums to infinity

OCR P2 5.2.2 (d)

If we have a look at the series

$$1 + \frac{1}{2} + \frac{1}{4} + \frac{1}{8} + \frac{1}{16} + \frac{1}{32} + \dots$$

we can see that the sum of this series to

1 term,	S_1	$=$	1
2 terms,	S_2	$=$	$1\frac{1}{2}$
3 terms,	S_3	$=$	$1\frac{3}{4}$
4 terms,	S_4	$=$	$1\frac{7}{8}$
5 terms,	S_5	$=$	$1\frac{15}{16}$
6 terms,	S_6	$=$	$1\frac{31}{32}$

and so on, with the sum getting closer and closer to 2. If we use our formula for the sum to n terms, since $a = 1$ and $r = \frac{1}{2}$ here:

$$S_n = \frac{1\left(1-\left(\frac{1}{2}\right)^n\right)}{1-\frac{1}{2}} = \frac{1-\left(\frac{1}{2}\right)^n}{\frac{1}{2}} = 2-\left(\frac{1}{2}\right)^{n-1}$$

So the only difference between S_n and 2 is the term $\left(\frac{1}{2}\right)^{n-1}$; and as n gets very large, this term gets very small. We can make S_n as close as we like to 2 by choosing a suitably large value of n. In situations like this, we say that the *limit* of the sum of the series as n tends to infinity is 2, or in other words, the series *converges* with a limit of 2, and write:

$$S_\infty = 2 \qquad \text{(or } S = 2\text{)}$$

You can experiment yourself with your calculator, taking any number between −1 and 1 and raising it to a large power. No matter how close the number is to +1 or −1, if the power is large enough the result is very small. For example:

$$0.999999^{99999999} \approx 3 \times 10^{-44}$$

which, by most standards, would be considered small. This fact becomes significant when we look at geometric series whose common ratio r is between +1 and −1. Their sum to n terms, which we have seen is:

$$S_n = \frac{a(1-r^n)}{1-r}$$

simplifies as n becomes large. The second term in the bracket, $-r^n$, since r lies within the correct range of values, simply disappears, leaving the top line of $(a \times 1)$ or a, i.e.

$$S_\infty = \frac{a}{1-r}$$

(Putting $a = 1$, and $r = \frac{1}{2}$ gives S_∞ of the series $1 + \frac{1}{2} + \frac{1}{4} + \frac{1}{8} + \dots$ the value of 2, which agrees with the value we found.)

Here are the results so far:

If u_n is the nth term of a GP,

$$u_n = ar^{n-1}$$

If S_n is the sum of the first n terms of a GP,

$$S_n = \frac{a(1-r^n)}{1-r}$$

If $-1 < r < 1$, this sum has a limit as n gets large, whose value is

$$S_\infty = \frac{a}{1-r}$$

We'll try applying these results to some examples.

Example

In a GP, the sum to infinity is 9 and the sum of the first two terms is 5. Find the first four terms of the progression, given that they are positive.

Solution

$$S_\infty = \frac{a}{1-r} = 9 \quad \Rightarrow a = 9(1-r) \qquad \text{①}$$

$$S_2 = a + ar = 5 \quad \Rightarrow a(1+r) = 5$$

$$\Rightarrow a = \frac{5}{1+r} \qquad \text{②}$$

Equating ① and ②, $9(1-r) = \dfrac{5}{1+r}$

$\Rightarrow 9(1-r)(1+r) = 5$

$9(1-r^2) = 5$

$1 - r^2 = \dfrac{5}{9}$ or $r^2 = \dfrac{4}{9}$

So $r = \pm\dfrac{2}{3}$, but the terms are positive, so $r = \dfrac{2}{3}$.

Putting this into ①, $a = 9\left(1 - \dfrac{2}{3}\right) = 3$

So the first four terms are $3,\ 3 \times \dfrac{2}{3},\ 3 \times \left(\dfrac{2}{3}\right)^2,\ 3 \times \left(\dfrac{2}{3}\right)^3$, i.e. $3,\ 2,\ \dfrac{4}{3},\ \dfrac{8}{9}$

Example

Find the range of values of x for which the geometric progression

$$1 - \left(\frac{x-1}{3}\right) + \left(\frac{x-1}{3}\right)^2 - \left(\frac{x-1}{3}\right)^3 + \ldots$$

has a sum to infinity.

Solution

The common ratio, $r = -\left(\dfrac{x-1}{3}\right) = \dfrac{1-x}{3}$

We need $\quad -1 < r < 1 \quad \Rightarrow \quad -1 < \dfrac{1-x}{3} < 1$

$\qquad -3 < 1 - x < 3$

$\qquad -4 < -x < 2 \qquad \times$ by -1

$\qquad 4 > x > -2 \quad \Rightarrow \quad -2 < x < 4.$

Example

The sum to infinity of a GP is 8 and the sum of the first three terms is 7. Find the first and fifth terms.

Solution

It's worth noting that the expression for the sum of the first n terms includes the expression for the sum to infinity

$$S_n = \frac{a(1-r^n)}{1-r} = \frac{a}{1-r} \times (1-r^n) = S_\infty(1-r^n)$$

Here we have $S_\infty = 8$ and $S_3 = 7$

$$S_3 = \frac{a(1-r^3)}{1-r} = S_\infty(1-r^3) = 8(1-r^3) = 7$$

$$\Rightarrow \quad 1 - r^3 = \frac{7}{8}, \quad r^3 = \frac{1}{8} \quad \Rightarrow \quad r = \frac{1}{2}$$

$$S_\infty = \frac{a}{1-r} = 8 \Rightarrow \frac{a}{1-\frac{1}{2}} = 8 \Rightarrow a = 4$$

Then $\quad a = 4$ and $u_5 = ar^4 = 4 \times \left(\dfrac{1}{2}\right)^4 = \dfrac{1}{4}$

| **Example** | A ball is dropped onto a horizontal plane and rebounds successively. The height above the plane reached by the ball after the first impact with the plane is 4 m. After each impact the ball rises to a height which is $\frac{3}{4}$ of the height reached after the previous impact. Calculate the total vertical distance travelled by the ball from the first impact until: |

(a)   the fourth impact

(b)   the eighth impact

(c)   it comes to rest.

| **Solution** | A diagram helps with examples like these: |

| **Figure 5.1** |

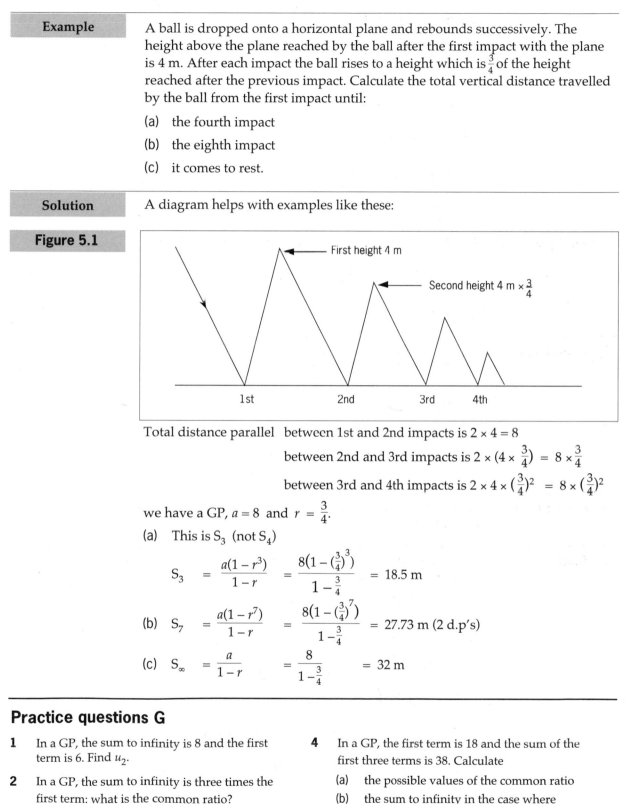

Total distance parallel   between 1st and 2nd impacts is $2 \times 4 = 8$

between 2nd and 3rd impacts is $2 \times (4 \times \frac{3}{4}) = 8 \times \frac{3}{4}$

between 3rd and 4th impacts is $2 \times 4 \times \left(\frac{3}{4}\right)^2 = 8 \times \left(\frac{3}{4}\right)^2$

we have a GP, $a = 8$ and $r = \frac{3}{4}$.

(a)   This is S_3 (not S_4)

$$S_3 = \frac{a(1 - r^3)}{1 - r} = \frac{8\left(1 - \left(\frac{3}{4}\right)^3\right)}{1 - \frac{3}{4}} = 18.5 \text{ m}$$

(b) $S_7 = \frac{a(1 - r^7)}{1 - r} = \frac{8\left(1 - \left(\frac{3}{4}\right)^7\right)}{1 - \frac{3}{4}} = 27.73 \text{ m (2 d.p's)}$

(c) $S_\infty = \frac{a}{1 - r} = \frac{8}{1 - \frac{3}{4}} = 32 \text{ m}$

Practice questions G

1   In a GP, the sum to infinity is 8 and the first term is 6. Find u_2.

2   In a GP, the sum to infinity is three times the first term: what is the common ratio?

3   In a GP, the sum to infinity is 4 and the second term is 1. Find u_1 and u_3.

4   In a GP, the first term is 18 and the sum of the first three terms is 38. Calculate

(a)   the possible values of the common ratio

(b)   the sum to infinity in the case where this exists.

5   Find the sum to n terms of the series

$$4 - 2x + x^2 - \frac{x^3}{2} + \dots$$

Find the range of values of x for which the sum to infinity exists.

6   The series S is defined by

$$S = 1 + 2x + 3x^2 + 4x^3 + \dots$$

Find an expression for $S - xS$, simplify the terms and hence deduce an expression for S.

7   The first term of a GP is 1 and the common ratio $r = \frac{1}{\sqrt{2}}$. Show that the sum to infinity of the series is $2 + \sqrt{2}$.

8   The first and second terms of a GP are 8 and 4. Show that the sum of all the terms after the nth is 2^{4-n}.

Applications

These series do crop up in practical situations and we can use the techniques to calculate particular sums or terms in which we're interested. One such situation is investment at a fixed rate of interest – we'll have a look in some detail at an example of this.

| **Example** | A 'Yearly Plan' is a National Savings scheme requiring 12 monthly payments of a fixed amount of money on the same date each month. All savings earn interest at a rate of $x\%$ per complete calendar month. |

A saver decides to invest £20 per month in this scheme and makes no withdrawals during the year. Show that, after 12 complete calendar months, his first payment has increased in value to:

$$£20r^{12}, \text{ where } r = 1 + \frac{x}{100},$$

Show that the total value, after 12 complete calendar months, of all 12 payments is:

$$£ \frac{20\,(r^{12} - 1)}{r - 1}.$$

Hence calculate the total interest received during the 12 months when the monthly rate of interest is $\frac{1}{2}$ per cent.

| **Solution** | After one month, the initial payment of £20 has a value of: |

$$£20 + £20 \times \frac{x}{100} \;=\; £20\left(1 + \frac{x}{100}\right)$$

$$= £20r$$

After a further month, this has increased to:

$$£20r + £20r \times \frac{x}{100} \;=\; £20r\left(1 + \frac{x}{100}\right)$$

$$= £20r^2$$

This continues, so that at the end of each month the value at the beginning of the month is increased by a factor of $1 + \frac{x}{100}$, i.e. r. Since after the first month, the value is £20r and after the second, £20r^2, the value at the end of the twelve months will be £20r^{12}.

For the second payment, at the beginning of the second month, the value will be £20r at the end of the month, £20r^2 at the end of the next month, the third and so on. At the end of the twelfth month its value will be £20r^{11}.

Similarly for the payments in each succeeding month: each has one less month to earn interest and the value at the end of the 12 months decreases by a factor of r each time, i.e. the third payment will be worth £20r^{10}, the fourth £20r^9 and so on until the twelfth payment, having only had one month in which to earn interest, is only worth £20r.

The total of these value will be:

$$£\left[\, 20r^{12} + 20r^{11} + 20r^{10} + \ldots + 20r \,\right]$$

If we turn this round for easier working, this becomes:

$$£\left[\, 20r + 20r^2 + \ldots + 20r^{11} + 20r^{12} \,\right]$$

This is a GP of twelve terms with first term £20r. We use the formula

$$S_n = \frac{a(r^n - 1)}{r-1} \text{ since } r > 1, \text{ and this gives:}$$

$$£\,\frac{20r(r^n - 1)}{r-1} \text{ as required.} \qquad\qquad [1]$$

When the monthly interest rate is $\frac{1}{2}\%$, i.e. $x - \frac{1}{2}$,

$$r = 1 + \frac{x}{100} = 1.005$$

Putting this value into equation [1] with $n = 12$ gives:

$$£\,\frac{20 \times 1.005\,[1.005^{12} - 1]}{1.005 - 1} = £247.94$$

The total interest received will be the difference between this sum and the total of the payments, i.e.

$$£247.94 - 12 \times £20 = £7.94$$

Practice questions H

1 A £100 investment in the 41st Issue of National Savings Certificates is worth £130.08 after 5 years. Is the Department for National Savings justified in saying that the investment grows at an annual rate of 5.4%? Carry out an appropriate calculation to justify your answer.

2 The value of a certain make of car when new is £13 500 and the value depreciates by 15% each year. Hence the value when the car is one year old is £13 500 × 0.85. Find the value of this make of car when it is eight years old, giving your answer to the nearest pound.

The value of a different make of car depreciates by 18% each year. Its value when new is £19 620. After n years its value is £1000. Write down an equation satisfied by n.

3 At the beginning of 1990, an investor decided to invest £6000 in a Personal Equity Plan (PEP), believing that the value of the investment should increase, on average, by 6% each year. Show that, if this percentage rate of increase is in fact maintained for 10 years, the value of the original investment will be about £10 745.

The investor added a further £6000 to the PEP at the beginning of each year between 1991 and 1995 inclusive. Assuming that the 6% annual rate of increase continues to apply, show that the total value, in £, of the PEP at the beginning of the year 2000 may be written as

$$6000 \sum_{r=5}^{10} (1.06)^r,$$

and evaluate this, correct to the nearest £.

SUMMARY EXERCISE

1 (a) The nth term of a sequence $u_1, u_2, u_3 \ldots$ is defined by the formula $u_n = 5 + (-1)^n$. Write down the first six terms of the sequence.

(b) The nth term of the oscillating sequence

$$8, 10, 8, 10, 8, 10, \ldots$$

is denoted by v_n. Write down a formula for v_n.

2 The nth term of a sequence is defined by $t_n = \frac{1}{2} n(n+1)$ for all positive integers n.

(a) Find the value of $t_1 + t_2$ and of $t_2 + t_3$.

(b) By simplifying an expression for $t_n + t_{n+1}$ show that the sum of any two consecutive terms is a perfect square.

3 The sequence $u_1, u_2, u_3, \ldots$ is defined by

$$u_n = 2n^2.$$

(a) Write down the value of u_3.

(b) Express $u_{n+1} - u_n$ in terms of n, simplifying your answer.

(c) The differences between successive terms of the sequence form an arithmetic progression. For this arithmetic progression, state its first term and its common difference, and find the sum of its first 1000 terms.

4 A sequence of numbers $u_1, u_2, \ldots, u_n, \ldots$ is given by the formula

$$u_n = 3 \left(\frac{2}{3} \right)^n - 1,$$

where n is a positive integer.

(a) Find the values of u_1, u_2 and u_3.

(b) Find $\displaystyle\sum_{n=1}^{15} 3 \left(\frac{2}{3} \right)^n$, and hence show that

$$\sum_{n=1}^{15} u_n = -9.014 \text{ to 4 significant figures.}$$

(c) Prove that $3u_{n+1} = 2u_n - 1$.

5 The first term of an arithmetic series is 3. The seventh term is twice the third term.

(a) Find the common difference

(b) Calculate the sum of the first 20 terms of the series. [AEB 1998]

6 In the arithmetic progression, the ninth term is 7, and the twenty-ninth term is equal to twice the fifth term.

(a) Determine the first term and the common difference of the progression.

(b) Calculate the sum of the first 200 terms of the progression. [AEB 1998]

7 A polygon has 10 sides. The lengths of the sides, starting with the smallest, form an arithmetic series. The perimeter of the polygon is 675 cm and the length of the longest side is twice that of the shortest side. Find, for this series,

(a) the common difference

(b) the first term.

8 A savings scheme pays 5% per annum compound interest. A deposit of £100 is invested in this scheme at the start of each year.

(a) Show that at the start of the third year, after the annual deposit has been made, the amount in the scheme is £315.25,

(b) Find the amount in the scheme at the start of the fortieth year, after the annual deposit has been made.

9 The first three terms of a geometric progression are $2x$, $x - 8$, $2x + 5$ respectively. Find the possible values of x.

[Hint: if a, b and c are consecutive terms of a GP, $\frac{b}{a} = \frac{c}{b}$.] [AEB 1997]

10 Find the sum of the first n terms of the sequence, $u_1, u_2, u_3, u_4 \ldots, u_r, \ldots$ in each of the following cases.

(a) $u_r = 3$ (b) $u_r = (-0.9)^r$

(c) $u_r = 1 + (-1)^r$ (d) $u_r = 3r + 4$.

11 A geometric series has third term 27 and sixth term 8.

(a) Show that the common ratio of the series is $\frac{2}{3}$

(b) Find the first term of the series.

(c) Find the sum to infinity of the series.

(d) Find, to 3 significant figures, the difference between the sum of the first 10 terms of the series and the sum to infinity of the series.

12 A competitor is running in a 25 km race. For the first 15 km, she runs at a steady rate of 12 km h^{-1}. After completing 15 km, she slows down and it is now observed that she takes 20% longer to complete each kilometre than she took to complete the previous kilometre.

(a) Find the time, in hours and minutes, the competitor takes to complete the first 16 km of the race.

The time taken to complete the rth kilometre is u_r hours.

(b) Show that, for $16 \le r \le 25$,
$$u_r = \frac{1}{12}(1.2)^{r-15}.$$

(c) Using the answer to (b), or otherwise, find the time, to the nearest minute, that she takes to complete the race.

SUMMARY

When you have finished this section, you should:

- be familiar with the use of u_n to denote the nth term of a sequence

- be able to work out particular terms of a sequence given a formula for the general term

- be able to find the general term of a given sequence in simple cases

- be familiar with the Σ-notation

- be able to write out particular terms and number of terms for a series given in Σ-form

- know the formula for an AP, $u_n = a + (n-1)d$

- know the formula for an AP, $S_n = \frac{n}{2}[2a + (n-1)d]$ and be able to prove it

- know the formula for a GP, $u_n = ar^{n-1}$

- know the formula for a GP, $S_n = \frac{a(1-r^n)}{1-r}$ and be able to prove it

- know the formula for a GP, $S_\infty = \frac{a}{1-r}$

- know that this exists only when $-1 < r < 1$

- be able to derive the nth term of a progression from the formula for S_n

- know how to apply the theory of GP's to applications involving interest and depreciation

- know that if the terms a, b and c are in

    (a) AP $\Rightarrow b - a = c - b$

    (b) GP $\Rightarrow \dfrac{b}{a} = \dfrac{c}{b}$

ANSWERS

Practice questions A

1 (a) 7, 9, 11, 13   (b) $\frac{1}{4}$, $\frac{1}{5}$, $\frac{1}{6}$, $\frac{1}{7}$

    (c) 3, 9, 27, 81   (d) $\frac{2}{3}$, $\frac{3}{5}$, $\frac{4}{6}$, $\frac{5}{7}$

    (e) 0, 2, 6, 12   (f) 6, 14, 24, 36

    (g) 2, 8, 24, 64   (h) 4, $\frac{8}{3}$, $\frac{16}{9}$, $\frac{32}{27}$

Practice questions B

1 (a) $3n + 1$ (b) $14 - 4n$

    (c) $\frac{n}{n+1}$ (d) $\frac{1}{2}(n+1)$

    (e) $n^2 - 1$ (f) n^3

Practice questions C

1 (a) 3, 5, 7 … 19, 21

    (b) 1, 3, 9 … 729, 2187

    (c) 30, 42, 56 … 132, 156

    (d) $\frac{4}{2}$, $\frac{5}{3}$, $\frac{6}{4}$ … $\frac{22}{20}$, $\frac{23}{21}$

    (e) 0, 2, 6 … 90, 110

    (f) 8, 24, 64 … 896, 2048

2 (a) 7   (b) 6   (c) 21   (d) $n + 1$

Practice questions D

1 290   **2** 1550   **3** 1683, 3367

4 9   **5** 6   **6** 23

7 –2   **8** 3000   **9** 82

10 $3\frac{1}{2}$, 1   **11** 12   **12** 5, –30; $k = 13$

13 600   **14** 15.2 cm   **15** 3

16 21

17 30°, 60°, 90°

18 $\frac{(n-m+1)}{2}(m+n)$

Practice questions E

1 4, 2   **2** 125

Practice questions F

1 512, 1023   **2** 2   **3** $\frac{4}{15}$

4 $\sqrt{12}$ **5** $\pm\frac{1}{2}$, $\pm\frac{1}{4}$ **6** 159.8 (1 dp)

7 1, $\frac{4}{9}$, 2.605 (3 d.p.)

8 7.549

Practice questions G

1 $\frac{3}{2}$ **2** $\frac{2}{3}$ **3** 2, $\frac{1}{2}$

4 (a) $\frac{2}{3}$, $-\frac{5}{3}$ (b) 54

5 $\dfrac{4(1 - (-\frac{x}{2})^n)}{1 + \frac{x}{2}}$, $-2 < x < 2$

6 $1 + x + x^2 + \dots$, $\dfrac{1}{(1-x)^2}$

Practice questions H

1 Yes, $100 \times 1.054^5 = $ £130.08 (nearest pence)

2 £3679, $19\,620 \times 0.82^n = 1000$

3 £56 007

Proof

INTRODUCTION You will find many questions start with the words: 'show that ...' or 'prove that ...' and the object of this section is to give you more idea about what kind of answer the examiner is looking for and how your solution should be structured. As an introduction to this rather different way of thinking, we will look at conditions for statements to be true: if p is true, does it follow that q is true, and does this work the other way round.

Necessary and sufficient conditions

It may help when deciding which type of condition is appropriate to think of them like this:

necessary	– maybe not enough
sufficient	– maybe too much
necessary and sufficient	– just right

Here's an example where the combination of two necessary conditions makes one necessary and sufficient condition.

Example

What conditions must be satisfied by a and b if $y = ax^2 + b$ is to be positive for all values of x.

Solution

We know that if the coefficient of x^2 is negative the graph of $y = ax^2 + b$ looks like this ...

Figure 6.1

$(0, b)$

$y = ax^2 + b$

and no matter how big the value of b, (so that the maximum point is high up on the y-axis), the curve will eventually come down and cross the x-axis, meaning that y is no longer positive. So we certainly have to have $a > 0$ if we want y to be always positive: in this case we say that $a > 0$ is a **necessary** condition.

But this in itself is not enough to ensure that the graph remains above the x-axis – we also need $b > 0$, which we can see if we look at the graph of $y = x^2 - 2$, for example

Figure 6.2

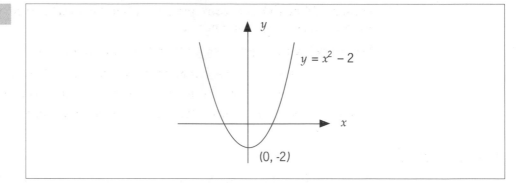

Similarly, we need $b > 0$ (so this is a necessary condition) but this in itself is not enough for y to be always positive (so it is not a **sufficient** condition) as we saw from the first graph.

However, if we have $a > 0$ *and* $b > 0$, the graph has a minimum point at $(0, b)$ and lies entirely above the x-axis

Figure 6.3

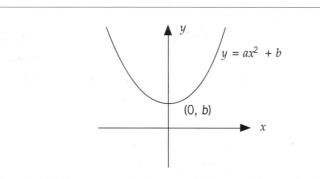

We say that a **necessary and sufficient** condition for $y = ax^2 + b$ to be positive for all x is that $a > 0$ and $b > 0$.

If we are after an example of a sufficient condition that is not a necessary condition we are looking for something that is *too much*. The condition that the number n has a factor of 6 is sufficient for a to have a factor of 3, but it is not a necessary condition: you can find a value of a such that it has a factor of 3 but not a factor of 6, e.g. $a = 9$.

Implication

The relationship between two statements can be written using the **implication sign** $\Rightarrow$. You have already seen this used in the context of worked examples where it has the meaning 'and therefore…' but here it has a slightly different meaning. The statement

$$p \Rightarrow q$$

means that if p is true, q is necessarily and always true. So for example we could write

$$x > 2 \Rightarrow x^2 > 4$$

and that is true: if $x > 2$ then always and invariably $x^2 > 4$. The reverse statement is not true …

$$x^2 > 4 \not\Rightarrow x > 2$$

because although it is sometimes true, it is not invariably true. $x^2 > 4$ is true for example when $x = -3$ and here $x \neq 2$.

We can relate this sign to the different types of condition.

$$p \Rightarrow q \quad \text{means that } p \text{ is a sufficient condition for } q$$
$$p \Leftarrow q \quad \text{means that } p \text{ is a necessary condition for } q$$
and finally $\quad p \Leftrightarrow q \quad$ means that p is a necessary and sufficient condition for q.

If and only if

A third way of describing the conditions is by using the terms **if**, **only if** and **if and only if** (which can be written as **iff**).

p is true if q is true	is the same as $q \Rightarrow p$
p is true only if q is true	is the same as $p \Rightarrow q$
p is true if and only if q is true	is the same as $p \Leftrightarrow q$

So for example:

p is a multiple of 2	if	p is a multiple of 6
p is a prime > 2	only if	p is odd
$x^3 = -8$	if and only if	$x = -2$

Example

Decide which of the following:

(i)   is a necessary condition for

(ii)   is a sufficient condition for

(iii) is a necessary and sufficient condition for

is appropriate to place between the following pairs of statements:

(a) p is even   ... pq is even (p, q positive integers)

(b) $x < 2$ ... $x^2 < 4$

(c) $x + 1 = \dfrac{4}{3}$ ... $\dfrac{1}{x} = 3$

Solution

(a) If p is even, then, whatever q is, pq is even (sufficient, i.e. $\Rightarrow$). But if pq is even, q could be even and p odd (not necessary, i.e. $\nLeftarrow$)

i.e. is a sufficient condition

(b) If $x^2 < 4$ then certainly $x < 2$ (necessary $\Leftarrow$).

But if $x < 2$, e.g. $x = -3$, $x^2 = 9$ so $x^2 \nless 4$ (not sufficient $\nRightarrow$)

i.e. is a necessary condition.

(c) Rearranging $x + 1 = \dfrac{4}{3}$, then $x = \dfrac{1}{3}$, then $\dfrac{1}{x} = 3$.

These are identical statements, each implying the other, $\Leftrightarrow$

i.e. is a necessary and sufficient condition.

Practice questions A

1   Between the following statements, insert
    either ⇒, ⇐ or ⇔ as appropriate

   (a) p and q odd ... pq odd

   (b) $x > 2 ... x^2 > 4$

   (c) $x = 0 ... xy = 0$

   (d) $x^2 = 9 ... x = 3$

   (e) p and q even ... $p + q$ even

   (f) $a^2 > b^2 ... a > b$

   (g) x is odd ... x^2 is odd

   (h) $pq > 0 ... \dfrac{p}{q} > 0$

   (i) $px^2 + qx + r > 0 ... p > 0$

   (j) $ax^2 + bx + c = 0$ has two real roots ...
       $b^2 > 4ac$

2   For each of the pair of statements in
    Question 1, insert one of the following:

   (a)   is a necessary condition for

   (b)   is a sufficient condition for

   (c)   is a necessary and sufficient condition
       for

3   Finally, repeat with if, only if, or if and only if.

Proof

A mathematical proof is a very particular structure, put together according to certain rules which became more and more strict as the level gets higher. In essence, it is similar to the game where you change one word into another in a series of steps, the rules being that you change one letter at a time and that the arrangement of letters at each change should be a recognisable word. For example, to change CAT into DOG, a possible route would be

$$\text{CAT} \quad \text{COT} \quad \text{COG} \quad \text{DOG}$$

Similarly, to show that two things are the same in mathematics you start at one end and work in steps until you reach the other: the requirement in this case being that each step relies on something which is generally accepted as being true or permissible.

Example

Prove the identity

$$\frac{1 + \sin x}{\cos x} + \frac{\cos x}{1 + \sin x} \equiv \frac{2}{\cos x}$$

Solution

Here we are going to start with the LHS and transform it step by step until we reach the RHS, hopefully.

$$\frac{1 + \sin x}{\cos x} + \frac{\cos x}{1 + \sin x} = \frac{(1 + \sin x)^2 + \cos^2 x}{\cos x \,(1 + \sin x)}$$

$$= \frac{1 + 2 \sin x + \sin^2 x + \cos^2 x}{\cos x \,(1 + \sin x)}$$

$$= \frac{1 + 2 \sin x + 1}{\cos x \,(1 + \sin x)}$$

$$= \frac{2 + 2 \sin x}{\cos x \,(1 + \sin x)}$$

$$= \frac{2(1 + \sin x)}{\cos x \,(1 + \sin x)}$$

$$= \frac{2}{\cos x}$$

All but one of the lines used a standard algebraic technique which were, respectively, gathering fractions together, expanding a bracket, collecting terms, factorising, cancelling. The exception was between lines 2 and 3, which used the fact that $\sin^2 x + \cos^2 x = 1$.

Note that in this case all the steps were reversible, so in fact we could have started from the RHS and worked our way to the LHS. This is not always so: we'll have a look at the proof of the factor theorem, which says that for a polynomial $f(x)$

$$(x - a) \text{ is a factor} \Leftrightarrow f(a) = 0$$

To work from left to right is quite easy. If $(x - a)$ is a factor, it means that $f(x) = (x - a)\,g(x)$ where $g(x)$ is the other factor: this is simply applying the meaning of factor. Then substituting $x = a$ gives

$$\begin{aligned} f(a) &= (a - a)\,g(a) \\ &= 0 \times g(a) \\ &= 0 \quad \text{as required} \end{aligned}$$

But this argument is not so easily reversed. If we start from the RHS, that $f(a) = 0$, where do we go! It's not so obvious just because a function is zero at a particular point that it should have factors, and this does not qualify as a generally accepted fact at our level (although it happens to be true).

So we have to be quite careful when we're working through, and particularly backwards, that we are justified in making each step. Here's a further example and you can have a go.

Example	Show that the equation $x^2 + px + q = 0$ has distinct real roots if, and only if, $p^2 > 4q$.

Solution	In cases like this, we are not entitled to use the quadratic formula and have to start from first principles. It's probably easiest to use a graphical argument. Starting with completing the square

$$y = x^2 + px + q = \left(x + \frac{p}{2}\right)^2 + \left(q - \frac{p^2}{4}\right)$$

this has a *minimum* point (since the x^2-coefficient is positive) at $\left(-\dfrac{p}{2},\, q - \dfrac{p^2}{4}\right)$

which we can call M. Now if M lies underneath the x-axis the curve has to cross the x-axis in two places, i.e. there are two solutions to the equation $y = x^2 + px + q = 0$.

Conversely, if the curve crosses the x-axis in two points, M lies below the axis.

Figure 6.4

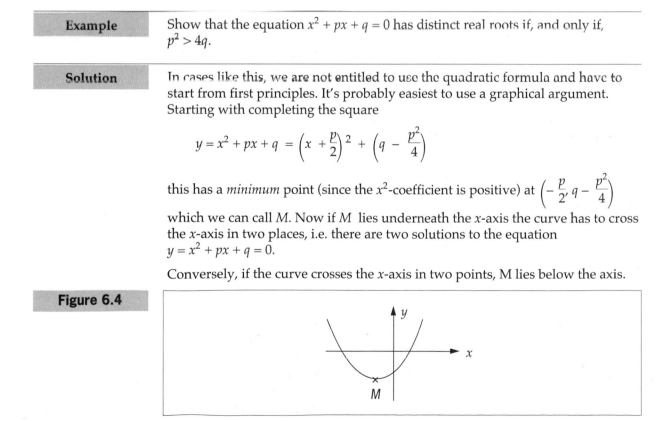

So if the y-coordinate of M is negative there are two distinct roots, i.e. if

$$q - \frac{p^2}{4} < 0 \iff \frac{4q - p^2}{4} < 0 \iff 4q - p^2 < 0$$

$$\iff p^2 > 4q$$

and we have shown that

$$x^2 + px + q = 0 \text{ has two distinct real roots} \iff p^2 > 4q.$$

Practice questions B

1 Show that for an arithmetic series

$$S_n = \frac{n}{2}\left[2a + (n-1)\,d\right]$$

2 Show that for a geometric series

$$S_n = \frac{a(1 - r^n)}{1 - r}$$

3 Prove the following trigonometric identities

(a) $\dfrac{\cos x}{\sin x} + \dfrac{\sin x}{\cos x} \equiv \dfrac{1}{\sin x \cos x}$

(b) $\dfrac{1}{\sin x} - \sin x \equiv \dfrac{\cos x}{\tan x}$

(c) $\dfrac{(\sin x + \cos x)^2}{\sin x \cos x} \equiv 2 + \dfrac{1}{\sin x \cos x}$

4 Show that x^2 is an odd integer if and only if x is an odd integer.

5 Show that if $ax^2 + bx + c = 0$,

$$x = \frac{-b \pm \sqrt{b^2 - 4ac}}{2a}$$

6 If $p = x + \dfrac{1}{x}$ and $q = x - \dfrac{1}{x}$, show that

$$p^2 - q^2 = 4$$

7 Show that for any integer a, $a^2 - a$ is even

8 Show that for any integer a, $a^3 - a$ is divisible by 6.

9 Show that for any acute angle θ,

$$\sin^2\theta + \cos^2\theta = 1$$

10 Show that if $x^2 > k(x + 1)$ for all x then $-4 < k < 0$

11 Show that $x^2 + 1 \geq 2x$ for any x.

12 Show that if $\dfrac{a}{b} = \dfrac{c}{d}$, then

(a) $\dfrac{a - c}{a + c} = \dfrac{b - d}{b + d}$

(b) $\dfrac{a - b}{a + b} = \dfrac{c - d}{c + d}$ (Tricky!)

SUMMARY   When you have finished this section, you should:

- know the difference between a necessary and a sufficient condition

- know what is meant by the statement that p is a necessary and sufficient condition for q

- be familiar with the symbols $\Rightarrow$, $\Leftarrow$ and $\Leftrightarrow$ and how these relate to the conditions above

- be familiar with the terms if, only if, and if and only if, and how these relate to the above conditions

- know what is meant by a mathematical proof

- be able to follow a proof line by line, appreciating the justification for each step

- be able to reproduce the standard proofs demanded in the specification

- be able to construct your own proof in simple cases, starting from one statement and finishing with the statement asked for in the question.

ANSWERS

Practice questions A

1. (a) $\Leftrightarrow$ (b) $\Rightarrow$ (c) $\Rightarrow$
   (d) $\Leftarrow$ (e) $\Rightarrow$ (f) $\Leftarrow$
   (g) $\Leftrightarrow$ (h) $\Leftrightarrow$ (i) $\Rightarrow$
   (j) $\Leftrightarrow$

2. (a) necessary and sufficient
   (b) sufficient
   (c) sufficient
   (d) necessary
   (e) sufficient
   (f) necessary
   (g) necessary and sufficient
   (h) necessary and sufficient
   (i) sufficient
   (j) necessary and sufficient

3. (a) iff   (b) only if
   (c) only if   (d) if
   (e) only if   (f) if
   (g) iff   (h) iff
   (i) only if   (j) iff

Practice questions B

1. Refer to Sequences and Series section.

2. Refer to Sequences and Series section.

3. (a) $\dfrac{\cos x}{\sin x} + \dfrac{\sin x}{\cos x} =$

   $\dfrac{\cos^2 x + \sin^2 x}{\sin x \cos x} = \dfrac{1}{\sin x \cos x}$

   (b) $\dfrac{1}{\sin x} - \sin x = \dfrac{1 - \sin^2 x}{\sin x}$

   $= \dfrac{\cos^2 x}{\sin x} = \dfrac{\cos x}{\tan x}$

   (c) $\dfrac{(\sin x + \cos x)^2}{\sin x \cos x}$

   $= \dfrac{\sin^2 x \cos^2 x + 2 \sin x \cos x}{\sin x \cos x}$

   $= \dfrac{1 + 2 \sin x \cos x}{\sin x \cos x} = 2 + \dfrac{1}{\sin x \cos x}$

4. If x is odd then x^2 is odd.
   If x is even then x^2 is even.
   So x is odd if x^2 is odd.

5. Refer to Algebra section.

6. $p^2 - q^2 = \left(x + \dfrac{1}{x}\right)^2 - \left(x - \dfrac{1}{x}\right)^2$

   $= x^2 + 2 + \dfrac{1}{x^2} - \left(x^2 - 2 + \dfrac{1}{x^2}\right) = 4$

7. $a^2 - a = a(a-1)$. Either a or $(a-1)$ is even, so product is even.

8. $a^3 - a = a(a^2 - 1) = a(a-1)(a+1) = (a-1)a(a+1)$

   Of these 3 consecutive numbers, one must be a multiple of 3 and one (at least) a multiple of 2 $\Rightarrow$ product must have a factor of 6.

9. Refer to Trigonometry section.

10. $x^2 > k(x+1) \Rightarrow x^2 - kx - k > 0$. No real roots,

    so $(-k)^2 - 4(1)(-k) > 0 \ \ k^2 + 4k > 0 \ \ k(k+4) > 0$

    $\Rightarrow -4 < k < 0$

11. $x^2 + 1 - 2x = x^2 - 2x + 1 = (x-1)^2 \geq 0$ for any
    $x \Rightarrow x^2 + 1 - 2x \geq 0 \Rightarrow x^2 + 1 \geq 2x$

12. (a) $\dfrac{(a-c)(b+d)}{(a+c)(b-d)} = \dfrac{ab + ad - bc - cd}{ab - ad + bc - cd}$.

    But if $\dfrac{a}{b} = \dfrac{c}{d} \Rightarrow ad = bc$

    $= \dfrac{ab - cd}{ab - cd} = 1 \Rightarrow \dfrac{a-c}{a+c} = \dfrac{b-d}{b+d}$.

    Similarly for (b)

Differentiation

Differentiation is a way of looking at how the change in one variable affects the change in a related variable. If we took two variables such as distance and time, we might study the change in distance as time changed: technically speaking this is the *rate of change of distance with respect to time*, or in other words, the speed.

From the equation of the relationship between the variables we can use differentiation to discover the associated rate of change. This has many applications once we have the basic array of techniques that allow us to differentiate certain algebraic expressions. Here we shall be looking at just two of these: finding the equations of tangents and normals to curves and maximising or minimising quantities subject to given constraints

The gradient of a curve

OCR P1 5.1.5 (a)

We saw in a previous section how the gradient of a straight line is defined by

$$\text{gradient} = \frac{y\text{-increase}}{x\text{-increase}}$$

Figure 7.1

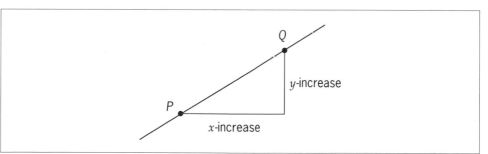

In the case of a straight line, this can be found quite easily – we take any two points, find the x– and y–increases and calculate the gradient, which is always the same. Once we know the gradient of a particular line, we can work out for any x–increase what the corresponding y–increase will be. If the gradient is 2 then an x–increase of 3 will mean a y–increase of 6, for example.

Things are different when we turn to different kinds of relationships between the variables x and y. Take the example of $y = x^2$. When we look at the graph we can see that the gradient is changing all the time. (0, 0), (1, 1) and (2, 4) are all on the curve but $\frac{y\text{-increase}}{x\text{-increase}}$ gives us $\frac{1}{1} = 1$ for the first two points and $\frac{3}{1} = 3$ for the next pair.

117

Figure 7.2

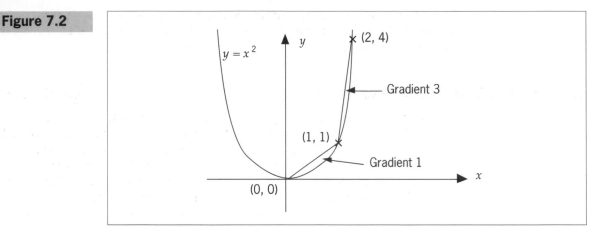

We have to find another way of defining the gradient which we can use for curves. The definition is

> The gradient of a curve at a particular point is defined to be the gradient of the tangent to the curve at that point.

Figure 7.3

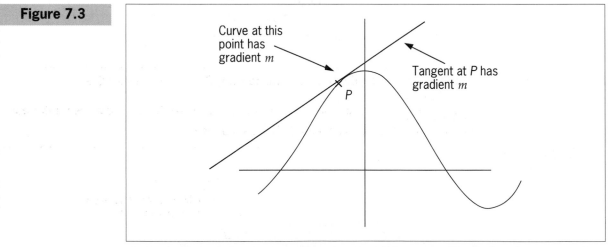

So in the figure above, since the tangent at P has a gradient of m, the *curve* at P has the same gradient.

Notation

We need some way of writing the gradient. Originally the y–increase was written δy and the x–increase δx, using the Greek letter delta giving the gradient as $\dfrac{\delta y}{\delta x}$. Then the ratio $\dfrac{\delta y}{\delta x}$ was subjected to a particular process which we shall look at shortly and the end result of that process was and is written $\dfrac{dy}{dx}$, the gradient of the curve at any point.

> The gradient of a curve is written $\dfrac{dy}{dx}$

We now know that the gradient of the curve is the same as the gradient of the corresponding tangent: the next step is to see how we can find the gradient of a tangent at any point.

The gradient of a tangent

OCR P1 5.1.5 (a)

To see how this works, we're going to take an actual example: finding the gradient of the curve $y = x^2$ at the point P where $x = 2$. To find this, we need to find the gradient of the tangent and we have no direct way of doing this: we have to approach it in a series of approximations. The first step is to take another point Q a little further along the curve, say when $x = 2.1$. Then since $y = x^2$, the corresponding y–coordinate will be $y = (2.1)^2 = 4.41$.

Figure 7.4

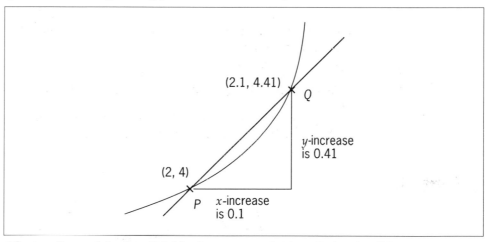

The gradient of the line PQ (the line joining PQ is called a *chord*) is

$\frac{4.41 - 4}{2.1 - 2} = \frac{0.41}{0.1} = 4.1$ and we can see that in this case this is an overestimate: the tangent at P has a gradient slightly less than this.

Figure 7.5

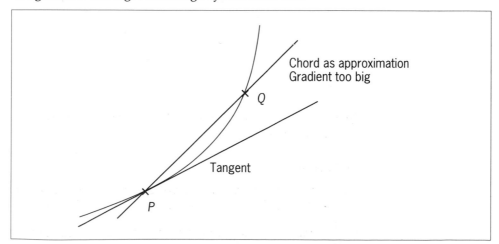

So we take another point R which we bring closer to P, where $x = 2.01$ for example. Then $y = (2.01)^2 = 4.0401$ and the gradient of the chord PR is

$$\frac{4.0401 - 4}{2.01 - 2} = \frac{0.0401}{0.01} = 4.01$$

This is a better approximation although still a slight overestimate.

So we take another point S, where $x = 2.001$. In this case the gradient turns out to be 4.001. If we put these results in a table we start to see a pattern emerging.

Table 7.1

	x–increase		gradient
PQ	2.1 – 2	= 0.1	4.1
PR	2.01 – 2	= 0.01	4.01
PS	2.001 – 2	= 0.001	4.001

It looks as though the gradient is 4 plus whatever the x–increase was. We can see why this is if we take a point P' slightly further on from $P(2, 4)$, say where $x = 2 + \delta$. Since $y = x^2$, the y–coordinate of P' will be $(2 + \delta)^2 = 4 + 4\delta + \delta^2$, so P' has coordinates $(2 + \delta, 4 + 4\delta + \delta^2)$. This gives a gradient of

$$\frac{4 + 4\delta + \delta^2 - 4}{2 + \delta - 2} \quad = \quad \frac{4\delta + \delta^2}{\delta} \quad = \quad 4 + \delta$$

which is 4 plus the increase, as we suspected. The point is that we are not restricted as to how close we take the second point, i.e. how small we make the increase δ. In fact, the closer we get to the original point P, the better the approximation to the gradient of the tangent. Since we can make δ as close as we like to zero we say that *in the limit* δ in fact disappears and we have the gradient *of the tangent* at P, which since $\delta = 0$ is just 4. It's hard to convince oneself that this reasoning is sound – how can the second point merge exactly with the first? You might like to look at Xeno's paradoxes, for example the one about the arrow never arriving, because it always has to travel half the remaining distance, and see whether this helps.

Tangent at any point

We have seen how the gradient of the curve $y = x^2$ at the point $x = 2$ is 4. If we want the gradient at a different point, say $x = 3$, we could repeat the whole process but in fact there is a way of finding the gradient at *any point* G on the curve. By supposing that the coordinates of G are (x, x^2) and the coordinates of a close point G' are $((x + \delta), (x + \delta)^2)$ we find the gradient is $\dfrac{2\delta x + \delta^2}{\delta} = 2x + \delta$.

Taking G' to be arbitrarily close to G so that $\delta \to 0$, we have the result that for $y = x^2$ the gradient is $2x$. In symbolic form, using our notation

$$\text{If } y = x^2 \quad \Rightarrow \quad \frac{dy}{dx} \quad = \quad 2x$$

You can see that it predicts the gradient at $x = 2$ to be $2 \times 2 = 4$, which is what we found. The gradient that we were looking for, at $x = 3$, is $2 \times 3 = 6$ and so on.

The gradient of $y = x^n$

OCR P1 5.1.5 (c)

In fact, we can extend this method and find the gradient function for $y = x^3$, for example. This turns out to be

$$\text{If } y = x^3 \text{, then the gradient function } \frac{dy}{dx} = 3x^2$$

We could repeat this procedure for any value of the power of x and we would find the general result that:

$$\text{The gradient function of the curve } y = x^n \text{ where } n \text{ is rational}$$
$$\text{is given by } \frac{dy}{dx} = nx^{n-1}$$

So for example,

if   (a) $y = x^5$, then $\dfrac{dy}{dx} = 5x^4$

       (b) $y = x^{-2}$, $\dfrac{dy}{dx} = -2x^{-3}$

       (c) $y = x^{\frac{1}{3}}$ $\dfrac{dy}{dx} = \dfrac{1}{3}x^{-\frac{2}{3}}$

If $y = f(x)$ where $f(x)$ is some function of x, a polynomial say, then we write the gradient function of $f(x)$ as $f'(x)$, i.e. $\frac{dy}{dx} = f'(x)$. [Similarly if $y = g(x)$, then $\frac{dy}{dx} = g'(x)$.]

$$\text{If } y = f(x) \text{ then the gradient function } \frac{dy}{dx} = f'(x)$$

This process of finding the gradient function is called *differentiation*. We differentiate $f(x)$ to give the *derivative* $f'(x)$.

Properties of gradients

OCR P1 5.1.5 (c)

1   If $y = f(x) + g(x) \Rightarrow \dfrac{dy}{dx} = f'(x) + g'(x)$

     e.g. if $y = x^4 + x^{-4}$ then $\dfrac{dy}{dx} = 4x^3 - 4x^{-5}$

     and if $y = x^7 - x^{-2}$ then $\dfrac{dy}{dx} = 7x^6 - (-2x^{-3}) = 7x^6 + 2x^{-3}$

2   If $y = af(x) \Rightarrow \dfrac{dy}{dx} = af'(x)$

     e.g. if $y = 5x^3 = 5(x^3)$

     then $\dfrac{dy}{dx} = 5(3x^2) = 15x^2$

     and if $y = -2x^{-\frac{1}{2}}$,

              $\dfrac{dy}{dx} = -2\left(-\dfrac{1}{2}x^{-\frac{3}{2}}\right) = x^{-\frac{3}{2}}$

3 If $y = a = ax^0$ (a constant) $\Rightarrow \dfrac{dy}{dx} = 0 \times \dfrac{d(x^0)}{dx} = 0$ i.e. constants on their own disappear.

e.g. $\quad y \quad = 6x^2 + \dfrac{6}{x^2} + 2 = 6x^2 + 6x^{-2} + 2$

then $\quad \dfrac{dy}{dx} \quad = 12x - 12x^{-3}$

You can probably see from the above examples that the method is to put the terms into the standard form ax^n and then use the general formula to find $\dfrac{dy}{dx}$.

Example

Differentiate with respect to x:

$\dfrac{6}{x^2}, \dfrac{5}{x^{\frac{1}{2}}}, \sqrt{x}, \sqrt[3]{x^2}$ i.e. find $\dfrac{dy}{dx}$)

Solution

$y \;=\; \dfrac{6}{x^2} \quad = 6x^{-2} \quad \Rightarrow \dfrac{dy}{dx} = -12x^{-3}$

$y \;=\; \dfrac{5}{x^{\frac{1}{2}}} \quad = 5x^{\frac{-1}{2}} \quad \Rightarrow \dfrac{dy}{dx} = -\dfrac{5}{2}x^{-\frac{3}{2}}$

$y \;=\; \sqrt{x} \quad = x^{\frac{1}{2}} \quad \Rightarrow \dfrac{dy}{dx} = \dfrac{1}{2}x^{-\frac{1}{2}}$

$y \;=\; \sqrt[3]{x^2} \quad = x^{\frac{2}{3}} \quad \Rightarrow \dfrac{dy}{dx} = \dfrac{2}{3}x^{-\frac{1}{3}}$

If we have to differentiate fairly simple products, like $y = (3x + 2)(x^2 - 1)$ or $\sqrt{x}\,(x^2 + 2)$, we multiply out first of all and then differentiate

$\quad y \quad = (3x + 2)(x^2 - 1) \qquad y = \sqrt{x}\,(x^2 + 2) \quad = x^{\frac{1}{2}}(x^2 + 2)$

$\quad\quad = 3x^3 + 2x^2 - 3x - 2 \qquad\qquad\qquad = x^{\frac{5}{2}} + 2x^{\frac{1}{2}}$

$\Rightarrow \dfrac{dy}{dx} = 9x^2 + 4x - 3 \qquad\qquad \Rightarrow \dfrac{dy}{dx} \quad = \dfrac{5}{2}x^{\frac{3}{2}} + x^{-\frac{1}{2}}$

Similarly if we have fractions where the bottom is a single term, we rewrite this as part of the top, multiply out and then differentiate.

Example

Find f'(x) when f$(x) = \dfrac{x^3 - 2x - 5}{3\sqrt{x}}$

Solution

$\dfrac{1}{3\sqrt{x}}$ is the same as $\dfrac{1}{3}x^{-\frac{1}{2}}$, so f$(x) = \dfrac{1}{3}x^{-\frac{1}{2}}(x^3 - 2x - 5)$

$= \dfrac{1}{3}x^{\frac{5}{2}} - \dfrac{2}{3}x^{\frac{1}{2}} - \dfrac{5}{3}x^{-\frac{1}{2}}$

$\Rightarrow$ f'$(x) \quad = \dfrac{1}{3} \times \dfrac{5}{2}x^{\frac{3}{2}} - \dfrac{2}{3} \times \dfrac{1}{2}x^{-\frac{1}{2}} - \dfrac{5}{3} \times -\dfrac{1}{2}x^{-\frac{3}{2}}$

$\quad\quad = \dfrac{5}{6}x^{\frac{3}{2}} - \dfrac{1}{3}x^{-\frac{1}{2}} + \dfrac{5}{6}x^{-\frac{3}{2}}$

Practice questions A

1 Find $\frac{dy}{dx}$ when $y =$

(a) x^3
(b) x^{10}
(c) x^7
(d) $x^{\frac{3}{4}}$
(e) $x^{\frac{2}{3}}$
(f) $x^{-\frac{4}{3}}$
(g) $\frac{1}{x}$
(h) $\sqrt{x}$
(i) $-\frac{1}{x^4}$
(j) $\frac{1}{\sqrt{x}}$
(k) $3x^4$
(l) $-\frac{5}{x^2}$
(m) $4\sqrt{x}$
(n) $\frac{1}{2x}$
(o) $\frac{3}{x^3}$
(p) $-\frac{4}{5}x^5$
(q) $\frac{2}{5x^5}$
(r) $(\sqrt{x})^3$
(s) $\left(\frac{1}{2x}\right)^2$
(t) $\sqrt{\frac{4}{x^3}}$

2 Find $f'(x)$ when $f(x) =$

(a) $x^2 + 6$
(b) $x^3 - x$
(c) $x^6 + 2x^3$
(d) $x^4 + x^2 + 1$
(e) $3x^4 - 5x^2 + 7$
(f) $5x^3 - 4x^2 + 7x - 9$
(g) $5x^2 + \frac{1}{5x^2}$
(h) $\frac{4}{x} - \frac{5}{x^3}$
(i) $2\sqrt{x} + \frac{1}{2\sqrt{x}}$
(j) $\frac{1}{x} + \frac{1}{2x^2} + \frac{1}{3x^3}$

3 Differentiate the following with respect to x:

(a) $(x + 1)^2$
(b) $(x - 4)(x + 4)$
(c) $(3x - 1)(x + 2)$
(d) $x(x^2 + 2)$
(e) $(2x - 3)^2$
(f) $x^{\frac{1}{2}}(x + 1)$
(g) $x^{-1}(x^2 + x + 1)$
(h) $x^{-\frac{1}{2}}(3x^2 + x - 5)$
(i) $x(x - 1)(x + 1)$
(j) $\sqrt{x}(4\sqrt{x} - 3)$

4 Express the following as the sum of terms of the form ax^k and hence differentiate with respect to x.

(a) $\frac{x^2 + 4}{x^2}$
(b) $\frac{3x - 2}{x}$
(c) $\frac{7x + 4}{3x}$
(d) $\frac{(x + 2)^2}{x}$
(e) $\frac{(2x - 3)^2}{x^2}$
(f) $\frac{(\sqrt{x} + 1)^2}{2x}$
(g) $\frac{3x^2 - 5x + 4}{3\sqrt{x}}$
(h) $\frac{(1 - x)(1 + x)}{x}$
(i) $\frac{(\sqrt{x} - 1)^2}{\sqrt{x}}$
(j) $\frac{(2\sqrt{x} - 1)^2}{(2\sqrt{x})^2}$

Tangents and normals

OCR P1 5.1.5 (d)

We found the gradient of the curve at any point by finding the gradient of the tangent at this point. We are now going to reverse this by finding the gradient of the curve (by differentiating) and deducing the gradient of the tangent. Knowing the gradient, we can find the equation of the tangent once we know the coordinates of the corresponding point. Let's see how this works with an example.

Example

Find the equation of the tangent to the curve whose equation is $y = x^2 - 2x + 3$ at the point where $x = 2$.

Solution

We differentiate the equation to find the gradient function

$$y = x^2 - 2x + 3 \implies \frac{dy}{dx} = 2x - 2$$

when $x = 2$, $\frac{dy}{dx} = 2 \times 2 - 2 = 2$ so that the gradient of the tangent is 2.

We put $x = 2$ into the original equation to find $y = 2^2 - 2 \times 2 + 3 = 3$. So we need the equation of a straight line with a gradient of 2 which passes through $(2, 3)$. This is

$$y - 3 = 2(x - 2) \implies y - 3 = 2x - 4$$
$$y = 2x - 1$$

This is the equation of the tangent at (2, 3). The *normal* at (2, 3) is the line which is perpendicular to the tangent

Figure 7.6

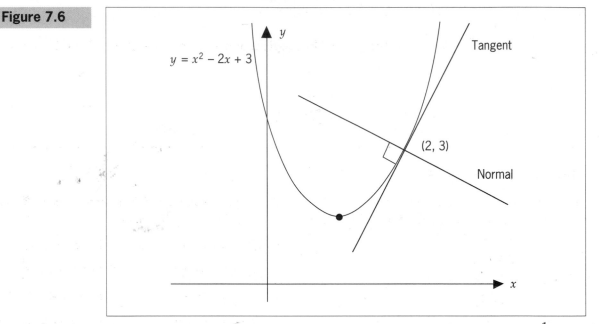

Since the gradient of the tangent is 2, the gradient of the normal will be $-\frac{1}{2}$ and the equation will be:

$$y - 3 = -\frac{1}{2}(x - 2)$$

$$2y - 6 = -x + 2 \quad \Rightarrow \quad x + 2y = 8$$

In summary, the procedure for finding the equation of a tangent or normal to the curve $y = f(x)$ is as follows:

1   Differentiate y to find the gradient function $\frac{dy}{dx}$

2   Find the value of this at the particular point. Call this gradient m

3   Find the coordinates of the point. Call these (x_1, y_1)

4   Equation of the tangent is $y - y_1 = m(x - x_1)$

Equation of the normal is $y - y_1 = -\frac{1}{m}(x - x_1)$

One further example where we have to find the x–coordinate first of all.

Example

Find the equation of the normal to the curve $y = x^3 - x$ at the point where it crosses the positive x–axis.

Solution

When the curve crosses the x–axis, $y = 0$ and so

$$x^3 - x = 0 \Rightarrow x(x^2 - 1) = 0$$
$$x(x - 1)(x + 1) = 0$$

so $x = 0$ or 1 or –1.

Since x is positive, $x = 1$.

Now we can start the procedure.

$$y = x^3 - x \implies \frac{dy}{dx} = 3x^2 - 1$$

when $x = 1$, $\frac{dy}{dx} = 3 - 1 = 2 \, (= m)$

We know that $y = 0$. The gradient of the normal is $-\frac{1}{m}$, i.e. $-\frac{1}{2}$ and the equation is

$$\begin{aligned} y - 0 &= -\frac{1}{2}(x - 1) \\ \implies 2y &= -x + 1 \implies x + 2y = 1 \end{aligned}$$

Practice questions B

1 Find the gradient of the tangent to the following curves at the stated point. State also the gradient of the normal.

(a) $y = x^2 + 4x$ at $x = 2$

(b) $y = 3x - 5x^3$ at $x = 1$

(c) $y = 3 - x^2$ at $x = -3$

(d) $y = x + \frac{1}{x}$ at $x = \frac{1}{2}$

(e) $y = \frac{1}{x^2}$ at $x = -2$

(f) $y = 4 + \frac{1}{4x}$ at $x = -1$

(g) $y = 5x^3 - \frac{2}{x}$ at $x = 1$

(h) $y = 3x^2 - 5x + 4$ at $x = 0$

2 Find the coordinates of the point(s) on the following curves where the gradient has the stated value:

(a) $y = x^2$ gradient 4

(b) $y = 3x^2 - 6x + 4$ gradient 0

(c) $y = 4\sqrt{x}$ gradient 1

(d) $y = 3x + \frac{1}{x^2}$ gradient 1

(e) $y = x^3 - 6x^2 + 10x + 5$ gradient -2

(f) $y = 2x^3 + 6$ gradient 24

(g) $y = \frac{4}{x}$ gradient -1

(h) $y = x + \frac{1}{x}$ gradient $\frac{3}{4}$

3 Find the equation of the tangent to the following curves at the given point:

(a) $y = x^3 - 8x^2 + 15x$ at the point $(4, -4)$

(b) $y = 2x^2 - 1$ at the point $(\frac{1}{2}, -\frac{1}{2})$

(c) $y = 4x + \frac{8}{x}$ at the point $(2, 12)$

(d) $y = 6x + \frac{2}{x^2} + 1$ at the point where $x = 1$

(e) $y = x^3 - x^2 - 6x$ at the point where $x = -1$

(f) $y = 5x - \frac{4}{x}$ at the point where $x = 1$

(g) $y = 3x^2 - 5x + 2$ at the point where it crosses the y-axis

(h) $y = \sqrt{x}\,(x - 2)$ at the point where $x = 4$

4 Find the equation of the normal of the following curves at the given point:

(a) $y = 2x^2 - 5x + 3$ at the point $(2, 1)$

(b) $y = 4\sqrt{x}$ at the point $(1, 4)$

(c) $y = (2x - 3)^2$ at the point where $x = 2$

(d) $y = x^2 - \frac{5}{x}$ at the point where $x = 1$

(e) $y = \frac{x + 1}{\sqrt{x}}$ at the point where $x = 4$

(f) $y = \frac{2x - 1}{x}$ at the point where it crosses the x-axis

(g) $y = \frac{(2\sqrt{x} + 1)^2}{3\sqrt{x}}$ at the point where $x = 1$

(h) $y = 4 - \frac{5}{\sqrt{x}}$ at the point where $x = 4$

Maximum and minimum points

OCR P1 5.1.5 (d)

As we've seen, the gradient of a curve $y = f(x)$ is written $\frac{dy}{dx}$ or $f'(x)$. It tells us what is happening to y as x increases: if the gradient is positive, y is increasing, if negative, y is decreasing. Let's have a look at part of a curve with tangents drawn in at different points …

Figure 7.7

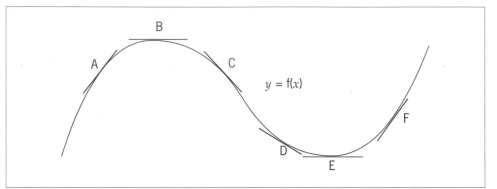

At the point A, the gradient is positive and at C it is negative. At some point in between, B, the gradient is *zero*. This is called a *maximum point* (although only a local maximum – there may be other regions where the value of the function is greater).

Similarly, at D the gradient is still negative, at F positive and at some point E, the gradient is zero: this is a *minimum point*.

> At maximum or minimum points, the gradient is zero, i.e. $\frac{dy}{dx} = 0$

Distinguishing between maximum and minimum

Anything in the form $\frac{d\square}{dx}$ means how $\square$ is changing as x increases. $\frac{dy}{dx}$ means how y is changing as x increases and $\frac{d\left(\frac{dy}{dx}\right)}{dx}$ means how $\frac{dy}{dx}$ or the gradient is changing as x increases. Since $\frac{d\left(\frac{dy}{dx}\right)}{dx}$ looks a bit cumbersome, it is written $\frac{d^2y}{dx^2}$, or $f''(x)$ if $y = f(x)$ and $\frac{dy}{dx} = f'(x)$.

> $\frac{dy}{dx}$ means how y is changing as x increases
>
> $\frac{d^2y}{dx^2}$ means how $\frac{dy}{dx}$ is changing as x increases

Now let's have a look back at our curve and see how the gradient changes as x increases:

A gradient positive
B gradient zero   } gradient is decreasing as x increases, i.e. $\dfrac{d^2y}{dx^2} < 0$
C gradient negative

D gradient negative
E gradient zero   } gradient is increasing as x increases, i.e. $\dfrac{d^2y}{dx^2} > 0$
F gradient positive

$$\frac{d^2y}{dx^2} < 0 \implies \text{a maximum point}$$

$$\frac{d^2y}{dx^2} > 0 \implies \text{a minimum point}$$

We now have a method for finding maximum and minimum points and distinguishing between them.

1   Given $y = f(x)$, we differentiate to find $\dfrac{dy}{dx}$.

2   We put $\dfrac{dy}{dx} = 0$ and solve this equation for x.

3   We put the value of x into $f(x)$ to find the y coordinate.

4   We differentiate $\dfrac{dy}{dx}$ to find $\dfrac{d^2y}{dx^2}$.

5   We put the value from (2) into $\dfrac{d^2y}{dx^2}$: if this is negative, the point is maximum: if positive, minimum.

Let's see how this works ...

| **Example** | Find the coordinates of the turning points for the curve: |

$$y = x^3 - 3x$$

| **Solution** | To find the gradient at any point, we differentiate: |

$$\frac{dy}{dx} = 3x^2 - 3$$

We want to find the points where the gradient is 0, i.e.:

$$3x^2 - 3 = 0, \qquad 3(x^2 - 1) = 0, \quad x^2 - 1 = 0$$

Then $x^2 = 1 \implies x = \pm1$ ← x–coordinates of turning points

When $x = 1$, $y = 1 - 3 = -2$

   $x = -1$, $y = -1 + 3 = 2$

So the two turning points are $(1, -2)$ and $(-1, 2)$.

Now differentiating $\dfrac{dy}{dx}$ gives:

$$\dfrac{d^2y}{dx^2} = 6x$$

when $x = 1$, $\quad \dfrac{d^2y}{dx^2} = 6 > 0 \quad\quad \Rightarrow \quad$ minimum

$ x = -1$, $\quad \dfrac{d^2y}{dx^2} = -6 < 0 \quad \Rightarrow \quad$ maximum

So the turning points are (1, –2) minimum and (–1, 2) maximum.

Another method of determining whether a point is a maximum or minimum (or neither), is to find the sign of the gradient on either side of the point in question. If the gradient is positive just before the point and negative just after.

Figure 7.8

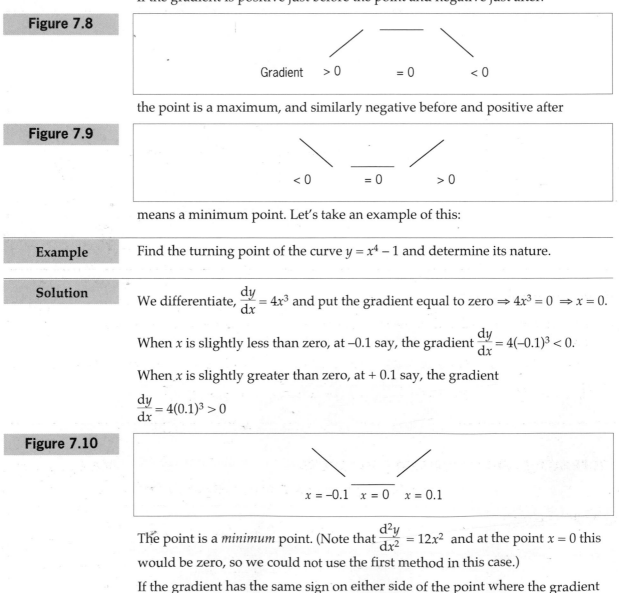

Gradient $\quad > 0 \quad\quad\quad = 0 \quad\quad\quad < 0$

the point is a maximum, and similarly negative before and positive after

Figure 7.9

$< 0 \quad\quad\quad = 0 \quad\quad\quad > 0$

means a minimum point. Let's take an example of this:

Example

Find the turning point of the curve $y = x^4 - 1$ and determine its nature.

Solution

We differentiate, $\dfrac{dy}{dx} = 4x^3$ and put the gradient equal to zero $\Rightarrow 4x^3 = 0 \Rightarrow x = 0$.

When x is slightly less than zero, at –0.1 say, the gradient $\dfrac{dy}{dx} = 4(-0.1)^3 < 0$.

When x is slightly greater than zero, at + 0.1 say, the gradient

$$\dfrac{dy}{dx} = 4(0.1)^3 > 0$$

Figure 7.10

$x = -0.1 \quad x = 0 \quad x = 0.1$

The point is a *minimum* point. (Note that $\dfrac{d^2y}{dx^2} = 12x^2$ and at the point $x = 0$ this would be zero, so we could not use the first method in this case.)

If the gradient has the same sign on either side of the point where the gradient is zero (the *stationary* point), the point is called a *point of inflection*.

Figure 7.11

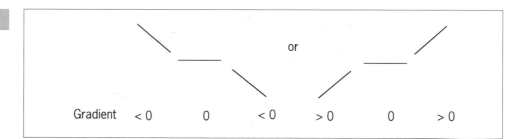

For example, the curve $y = x^3$ has a stationary point when $x = 0$. The gradient

$\frac{dy}{dx} = 3x^2$ and this is positive either side of $x = 0$, so the stationary point is a point of inflection.

Figure 7.12

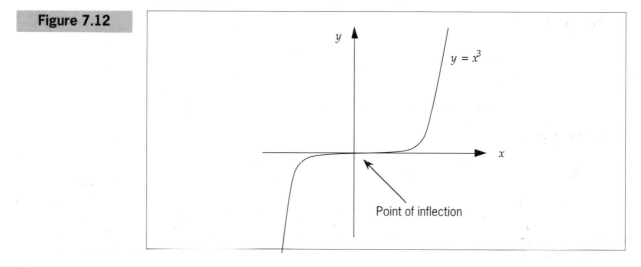

Point of inflection

Practice questions C

1   Find the stationary point(s) for the following curves. For each point, determine whether it is a maximum or minimum.

(a) $y = 2x^2 - 8x + 3$ (b) $y = x^3 - 12x$ (c) $y = x^2 + 1$ (d) $y = x + \frac{4}{x}$ (e) $y = x^4 + 4x + 1$

(f) $y = x^2 + \frac{16}{x}$ (g) $y = \frac{(x-1)^2}{x}$ (h) $y = \sqrt{x}\,(x - 8)$ (i) $y = (\sqrt{x} - 1)^2$

Increasing and decreasing functions

OCR P1 5.1.5 (d)

Some graphs have a positive gradient throughout like:

Figure 7.13

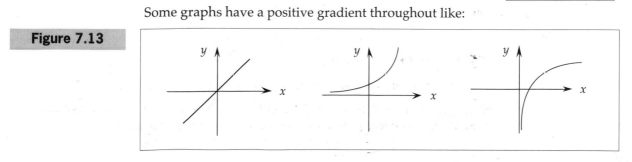

These are called *increasing functions*. Similarly, some graphs have a negative gradient throughout:

Figure 7.14

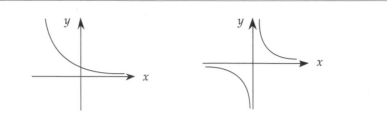

and these are called, as you may be able to guess, *decreasing* functions. They have no turning points, since the gradient is never zero.

Practice questions D

1   Determine whether the following are increasing, decreasing or neither by finding $\frac{dy}{dx}$ and seeing whether this is always positive, always negative or a mixture:

(a) $y = 2x + 3$ (b) $3x + 2y = 7$ (c) $y = \frac{4}{x}$ (d) $y = 5 - \frac{3}{2x}$ (e) $y = x^3 + 3$

(f) $y = x^2 + 3$ (g) $y = x + \frac{1}{x}$ (h) $y = 5x - \frac{3}{x}$ (i) $y = 3x^3 - \frac{3}{x^3}$ (j) $y = 4\sqrt{x} + 9$

Maximum and minimum problems   OCR P1 5.1.5 (d)

These problems could probably be summarised as 'making the most of available resources'. Suppose I have a piece of wire 40 cm long. If I make a rectangle from the wire, how should I choose the length and breadth to give me the greatest possible area? I could choose the length to be 15 cm and the breadth 5 cm.

Figure 7.15

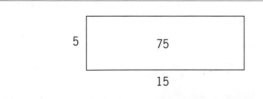

This gives me the correct perimeter of 40 cm: the area would be $5 \times 15 = 75$ cm^2.

Another possibility would be 8 cm by 12 cm.

Figure 7.16

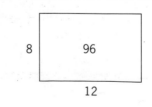

This would give an area of 96 cm^2, an improvement on the last attempt.

If I want to try and put the problem in mathematical terms, I need to give the variables names, so we can call the breadth x cm, the length y cm and the area A cm^2.

Figure 7.17

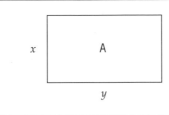

Now there are two quantities: one fixed, the perimeter, and the other varying, the area. We can express these as equations in x and y:

perimeter: $2x + 2y = 40$ [1]   FIXED
area: $xy = A$ [2]   VARYING

The strategy is to express the quantity that is varying in terms of only one variable. To do this, we need to make the other variable the subject of the fixed equation. In this case, it doesn't really matter which variable we take, but we'll take y:

$$2y = 40 - 2x$$
(divide by 2) $y = 20 - x$ [3]

we now substitute this expression into equation [2], which gives:

$$x(20 - x) = A$$

If we turn this round and multiply out the bracket:

$$A = 20x - x^2$$

we have now expressed the quantity that can vary in terms of just one of the variables.

We can now use our method for finding maximum and minimum values of a function – we differentiate our expression for A with respect to x and equate the result to zero.

$$\frac{dA}{dx} = 20 - 2x = 0$$
$$2x = 20$$
$$\Rightarrow x = 10$$ [4]

When we put this value back into equation [3], we find that

$$y = 20 - 10 = 10$$

So the rectangle which encloses the maximum area is a square of side 10 cm and this maximum area is 100 cm^2. (We could actually have enclosed a bigger area if we had used a circle with a perimeter of 40 cm: the area then works out to be a little more than 127 cm^2.)

Here is another example, a little more difficult this time, but you will see that the basic pattern of the solution is the same.

| Example | A solid right circular cylinder has a fixed volume of 1000 cm³. Show that the total surface area A cm² of the cylinder is related to the base radius x cm of the cylinder by the equation: |

$$A = 2\pi x^2 + \frac{2000}{x}$$

Given that x varies, show that A is a minimum when $x^3 = \frac{500}{\pi}$.

Find the minimum value of A, giving your answer to 1 decimal place.

| Solution | A right circular cylinder means that the sides are at right-angles to the base. In all these questions, it probably helps to draw a diagram and mark in some lengths: |

Figure 7.18

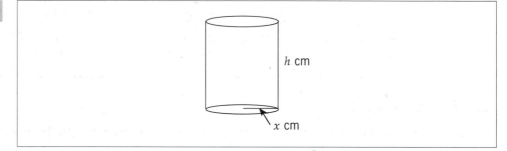

Since we're not told the height, we can call it h cm. We're looking for two quantities, one fixed and one varying. In this case the fixed quantity is the volume, which for a cylinder is $\pi x^2 h$ cm³ where x cm is the radius of the base. The quantity which varies, the surface area, is made up of two parts: $2\pi x h$ cm² for the curved surface and $2 \times \pi x^2$ cm² for the top and bottom. Our two equations are:

$$V \quad = \pi x^2 h \quad = 1000 \quad \text{(given)} \qquad [1] \qquad \text{FIXED}$$
$$A \quad = 2\pi x h + 2\pi x^2 \qquad\qquad\qquad [2] \qquad \text{VARYING}$$

We have a choice of making x or h the subject of equation [1]. It's easier to take h, particularly as h only occurs once in the second equation.

From [1] $h \quad = \dfrac{1000}{\pi x^2}$ \qquad\qquad\qquad [3]

and putting this into [2]

$$A \quad = 2\pi x \left(\frac{1000}{\pi x^2}\right) + 2\pi x^2 = \frac{2000}{x} + 2\pi x^2$$

i.e. $A \quad = 2\pi x^2 + \dfrac{2000}{x}$ \qquad\qquad [4]

We can now differentiate this with respect to x, remembering that we can re-write $\dfrac{2000}{x}$ as $2000\,x^{-1}$:

$$\frac{dA}{dx} = 4\pi x - \frac{2000}{x^2}$$

Putting this equal to zero:

$$4\pi x - \frac{2000}{x^2} = 0$$

$$4\pi x = \frac{2000}{x^2}$$

$$4\pi x^3 = 2000$$

$$x^3 = \frac{2000}{4\pi} = \frac{500}{\pi} \quad \text{as required.}$$

To show that this is a minimum, we need to find $\dfrac{d^2A}{dx^2}$ and show that it's positive when x has this value.

$$\frac{d^2A}{dx^2} = 4\pi + \frac{4000}{x^3}$$

Since $x > 0$, we can see this is positive without working out the exact value.

Using a calculator, we find that $x = 5.419$ (3 decimal places) and when we put this value into equation [4] we get A = 553.6, the required minimum value.

Practice questions E

1 For each of these questions, rearrange the first equation to make the variable y the subject. Substitute this into the second equation so that V is just in terms of x. Then follow the procedure outlined in the section on distinguishing maximum and minimum points, to find:

- the value of x for which $\dfrac{dV}{dx} = 0$
- the corresponding value of V
- whether this is a maximum or a minimum value.

(a) $x + y = 4$ $V = x^2 + 3xy$ (b) $2x + y = 10$ $V = 3x^2 - xy$

(c) $xy = 3$ $V = y(1 + 4x^3)$ (d) $xy = 4$ $V = y^2 + \dfrac{x^4y}{12}$

SUMMARY EXERCISE

1 (a) Given that $y = \dfrac{x^2 + 2}{x}$, find $\dfrac{dy}{dx}$

   (b) Find the gradient of the curve $y = \dfrac{1}{x^2}$ at the point where $x = 2$

2 P is the point (4, 7) on the curve $y = x^2 - 6x + 15$. Find the gradient of the curve at P, and the equation of the tangent at this point.

The tangent at another point Q is perpendicular to the tangent at P. Calculate the x-coordinate of Q.

3 The gradient of the curve $y = ax^2 + bx$ at the point (2, 4) is –8. Find two equations connecting a and b and hence find the values of a and b.

4 A curve has the equation $y = 2x^2 - 5x + 3$. Find:

   (a) the x-coordinate of the minimum points

   (b) the equation of the normal to the curve at the point where $x = 2$.

5 The equation of a curve is $y = 4x + \dfrac{1}{x}$. Find:

   (a) the coordinates of the turning points

   (b) the equation of the tangent to the curve at the point where $x = 2$.

6 The function f is defined for all real values of x by:

$$f(x) = x^3 - 6x^2 + 13x - 5.$$

(a) Differentiate $f(x)$ with respect to x to obtain $f'(x)$.

(b) Express $f'(x)$ in the form $a(x - b)^2 + c$, where a, b and c are constants whose values should be stated. Hence show that f is an increasing function.   [AQA 1999]

7 Find the equation of the tangent to the curve with equation

$$y = x^2 - \frac{2}{x}$$

at the point $P(1, -1)$.

Determine the coordinates of the point where the tangent at P intersects the curve again.

8 For a particular journey of a ship, the running cost, C, in hundreds of pounds, is given in terms of its average speed for the journey, v km h^{-1}, by the equation

$$C = \frac{16\,000}{v} + v^2.$$

(a) Use differentiation to calculate the value of v for which C is a minimum.

(b) Show that C is a minimum and not a maximum for this value of v.

9 A fluid flows along a straight shallow channel with parallel sides. At a point x cm from one side of the channel, the speed, v cms^{-1}, of the fluid is given by

$$v = 3.4 + 0.051x - 0.0003x^2.$$

(a) Find by differentiation the maximum speed of the fluid as x varies, giving your answer correct to 2 significant figures.

(b) Assuming that the speed of the fluid is greatest at the point half-way across the channel, find the width of the channel.

10 The expected sales, S millions, of an item when it is priced at £P, can be modelled by the equation

$$S = 140 - 3P^2, \quad 0 < P < 6.$$

The total revenue, £R million, is given by

$$R = PS.$$

Express R in terms of P and hence find $\frac{dR}{dP}$.

Calculate the value of P, to the nearest penny, required to maximise the total revenue, and find the value of this total revenue.

11 The diagram shows a rectangular box with a square base of side x cm and a height of $(20 - x)$ cm. Write down, in terms of x, an expression for the volume of the box.

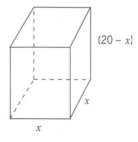

Find, by differentiation, the value of x which makes this volume a maximum and verify that, for this value of x, the volume is a maximum and not a minimum.

12 A large tank in the shape of a cuboid is to be made from 54 m^2 of sheet metal. The tank has a horizontal rectangular base and no top. The height of the tank is x metres. Two of the opposite vertical faces are squares.

(a) Show that the volume, V m^3, of the tank is given by

$$V = 18x - \frac{2}{3}x^3.$$

(b) Given that x can vary, use differentiation to find the maximum value of V.

(c) Justify that the value of V you have found is a maximum.

13

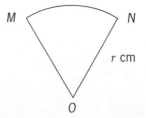

The figure shows a minor sector OMN of a circle centre O and radius r cm. The perimeter of the sector is 100 cm and the area of the sector is A cm^2.

(a) Show that $A = 50r - r^2$.

Given that r varies, find:

(b) the value of r for which A is a maximum and show that A is a maximum

(c) The value of $\angle MON$ for this maximum area

(d) the maximum area of the sector OMN.

14 A glass window consists of a rectangle with sides of length 2*r* cm and *h* cm and a semicircle of radius *r* cm. The total area of one surface of the glass is 500 cm².

2*r*

(a) (i) Write down a formula connecting *h* and *r*.

(ii) The perimeter of the window is *p* cm. By eliminating *h*, show that
$$p = \left(2 + \frac{\pi}{2}\right) r + \frac{500}{r}.$$

(b) (i) Determine the positive value of *r* for which *p* has a stationary value, giving your answer correct to three significant figures.

(ii) Calculate $\dfrac{d^2 p}{dr^2}$ and hence determine whether this stationary value is a maximum or minimum value.

15

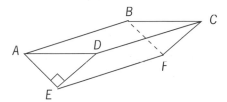

The figure above shows an open tank for storing water, *ABCDEF*. The sides *ABFE* and *CDEF* and rectangles. The triangular ends, *ADE* and *BCF* are isosceles and $\angle AED = \angle FCF = 90°$. The ends *ADE* and *BCF* are vertical and *EF* is horizontal. Given that *AD* = *x* metres.

(a) show that the area of Δ *ADE* is $\frac{1}{4} x^2$ m².

Given also that the capacity of the container is 4000 m³ and that the total area of the two triangular and two rectangular sides of the container is *S* m²

(b) show that $S = \dfrac{x^2}{2} + \dfrac{16\,000\sqrt{2}}{x}$.

Given that *x* can vary,

(c) use calculus to find the minimum value of *S*

(d) justify that the value of *S* you have found is a minimum.

SUMMARY

Now you have finished this section, you should:

- know that the gradient of a curve at any point is defined to be the gradient of the tangent at that point.

- know that the gradient function is written $\dfrac{dy}{dx}$ or f '(*x*)

- appreciate that the gradient of the tangent is found by a series of approximations by the gradients of associated chords

- know that if $y = x^n$, $\dfrac{dy}{dx} = nx^{n-1}$

- know how to differentiate sums and differences of algebraic terms

- know how to rearrange products and fractions so that they can be differentiated directly

- know how to find the equation of tangents and normals to the curve at any point

- know how to find stationary points on a curve

- know how to distinguish between maximum and minimum points by working out the sign of $\dfrac{d^2 y}{dx^2}$

- know that increasing functions have a positive gradient, and decreasing functions a negative gradient

- be able to solve practical problems involving a quantity to be maximised or minimised, subject to a fixed constraint.

ANSWERS

Practice questions A

1 (a) $3x^2$ (b) $10x^9$ (c) $7x^6$

(d) $\frac{3}{4}x^{-\frac{1}{4}}$ (e) $\frac{2}{3}x^{-\frac{1}{3}}$ (f) $\frac{-4}{3}x^{-\frac{7}{3}}$

(g) $\frac{-1}{x^2}$ (h) $\frac{1}{2\sqrt{x}}$ (i) $\frac{4}{x^5}$

(j) $\frac{-1}{2x^{\frac{3}{2}}}$ (k) $12x^3$ (l) $\frac{10}{x^3}$

(m) $\frac{2}{\sqrt{x}}$ (n) $\frac{-1}{2x^2}$ (o) $\frac{-9}{x^4}$

(p) $-4x^4$ (q) $\frac{-2}{x^6}$ (r) $\frac{3}{2}\sqrt{x}$

(s) $\frac{-1}{2x^3}$ (t) $-3x^{-\frac{5}{2}}$

2 (a) $2x^2$ (b) $3x^2-1$ (c) $6x^5+6x^2$

(d) $4x^3+2x$ (e) $12x^3-10x$

(f) $15x^2-8x+7$

(g) $10x-\frac{2}{5x^3}$ (h) $\frac{-4}{x^2}+\frac{15}{x^4}$

(i) $\frac{1}{\sqrt{x}}-\frac{1}{4x^{\frac{3}{2}}}$ (j) $\frac{-1}{x^2}-\frac{1}{x^3}-\frac{1}{x^4}$

3 (a) $2x+2$ (b) $2x$ (c) $6x+5$

(d) $3x^2+2$ (e) $8x-12$

(f) $\frac{3}{2}x^{\frac{1}{2}}+\frac{1}{2}x^{-\frac{1}{2}}$ (g) $1-x^{-2}$

(h) $\frac{9}{2}x^{\frac{1}{2}}+\frac{1}{2}x^{-\frac{1}{2}}+\frac{5}{2}x^{-\frac{3}{2}}$

(i) $3x^2-1$ (j) $4-\frac{3}{2}x^{-\frac{1}{2}}$

4 (a) $1+4x^{-2}:-8x^{-3}$ (b) $3-2x^{-1}:2x^{-2}$

(c) $\frac{7}{3}+\frac{4}{3}x^{-1}:\frac{-4}{3}x^{-2}$

(d) $x+4+4x^{-1}:1-4x^{-2}$

(e) $4-12x^{-1}+9x^{-2}:12x^{-2}-18x^{-3}$

(f) $\frac{1}{2}+x^{-\frac{1}{2}}+\frac{1}{2}x^{-1}:-\frac{1}{2}x^{-\frac{3}{2}}-\frac{1}{2}x^{-2}$

(g) $x^{\frac{3}{2}}-\frac{5}{3}x^{\frac{1}{2}}+\frac{4}{3}x^{-\frac{1}{2}}:\frac{3}{2}x^{\frac{1}{2}}-\frac{5}{6}x^{-\frac{1}{2}}-\frac{2}{3}x^{-\frac{3}{2}}$

(h) $x^{-1}-x:-x^{-2}-1$

(i) $x^{\frac{1}{2}}-2+x^{-\frac{1}{2}}:\frac{1}{2}x^{-\frac{1}{2}}-\frac{1}{2}x^{-\frac{3}{2}}$

(j) $1-x^{-\frac{1}{2}}+\frac{1}{4}x^{-1}:\frac{1}{2}x^{-\frac{3}{2}}-\frac{1}{4}x^{-2}$

Practice questions B

1 (a) $8,-\frac{1}{8}$ (b) $-12,\frac{1}{12}$ (c) $6,-\frac{1}{6}$

(d) $-3,\frac{1}{3}$ (e) $\frac{1}{4},-4$ (f) $-\frac{1}{4},4$

(g) $17,-\frac{1}{17}$ (h) $-5,\frac{1}{5}$

2 (a) $(2,4)$ (b) $(1,1)$ (c) $(4,8)$

(d) $(1,4)$ (e) $(2,9)$ (f) $(2,22),(-2,-10)$

(g) $(2,2),(-2,-2)$ (h) $(2,\frac{5}{2}),(-2,\frac{-5}{2})$

3 (a) $x+y=0$ (b) $2y=4x-3$

(c) $y=2x+8$ (d) $y=2x+7$

(e) $x+y=3$ (f) $y=9x-8$

(g) $5x+y=2$ (h) $2y=5x-12$

4 (a) $x+3y=5$ (b) $x+2y=9$

(c) $x+4y=6$ (d) $x+7y+27=0$

(e) $32x+6y=143$ (f) $2x+8y=1$

(g) $2x+y=5$ (h) $32x+10y=143$

Practice questions C

1 (a) $(2,-5)$ MIN

(b) $(-2,16)$ MAX $(2,-16)$ MIN

(c) $(0,1)$ MIN (d) $(2,4)$ MIN $(-2,-4)$ MAX

(e) $(-1,-2)$ MIN (f) $(2,12)$ MIN

(g) $(1,0)$ MIN $(-1,-4)$ MAX

(h) $\left(\frac{8}{3},-\frac{16}{3}\sqrt{\frac{8}{3}}\right)$ MIN (i) $(1,0)$ MIN

Practice questions D

1 (a) INC (b) DEC (c) DEC (d) INC

(e) INC (f) neither (g) neither

(h) INC (i) INC (j) INC

Practice questions E

1 (a) $x=3, V=18$ MAX (b) $x=1, V=-5$ MIN

(c) $x=\frac{1}{2}, V=9$ MIN (d) $x=2, V=\frac{20}{3}$ MIN

Integration

Integration is an enormous subject and each unit that you study goes a little further on from what went before, so a good grasp of the basics is very important. In this first unit, P1, we look at some of the fundamentals: the general formula for integrating a term of the form x^n and how we can apply this to more complicated algebraic expressions. We move into an application of integration: finding a function given some information about how it is changing.

We already know how to find the area enclosed by shapes with straight lines as sides: triangles, trapezia, etc. The final point of the section is devoted to using integration to find the area of shapes whose sides include some curves.

The reverse of differentiation $\boxed{\textbf{OCR P1}\ 5.1.6\ (a)}$

We can think of $\int 3x^2\,dx$ as meaning – 'what function do I have to differentiate with respect to x to give $3x^2$?' The answer, as you can probably work out, would be x^3.

If you can remember, we found a general result for differentiation, that

$$\frac{d(x^n)}{dx} = nx^{n-1}$$

so that, for example, $\dfrac{d(x^7)}{dx} = 7x^6$ and $\dfrac{d(x^{-1/2})}{dx} = -\dfrac{1}{2}x^{-3/2}$

We'll try and find a similar type of formula for integration, starting by working out a few individual results.

Suppose that we want to work out $\int x^2\,dx$

which means finding a function that would give x^2 after differentiation.

Since x^3 differentiates to $3x^2$, $\frac{1}{3} \times x^2$ differentiates to $\frac{1}{3} \times 3x^2 = x^2$, so $\dfrac{x^2}{3}$ differentiates to x^2.

Now let's try to find a function which gives x^4 after differentiation:

i.e. $\int x^4\,dx$

Again we start at one power higher : $\dfrac{d(x^5)}{dx} = 5x^4$, and we need to adjust the constant. Dividing by 5 gives $\dfrac{d(\frac{1}{5}x^5)}{dx} = \frac{1}{5} \times 5x^4 = x^4$, which is now correct.

General formula

From the above examples you can maybe begin to see a pattern. If we want to find a function which differentiates to x^n, we have to increase the power by 1 and divide the result by the new power, i.e. we differentiate $\dfrac{x^{n+1}}{n+1}$.

You can check that this formula gives you the results that you've already found. It seems to work well – but there's something missing. We could in fact differentiate any of $x^3 + 4$, $x^3 - 2$, $x^3 + \pi$, etc. and still come up with the answer $3x^2$ because any constant tacked onto the main function disappears when you differentiate. Without knowing anything more about particular values of the function, it's not possible to give an exact value to this constant, and so we write $\int x^2\,dx = \frac{1}{3}x^3 + C$ where C is an arbitrary constant.

This general result is written as:

$$\int x^n\,dx = \frac{x^{n+1}}{n+1} + C$$

$$(\text{provided } n \neq -1)$$

The exception is $\int \frac{1}{x}\,dx$. When we use the formula we end up with $\frac{x^0}{0}$: this doesn't make mathematical sense, but there is in fact a function which differentiates to give $\frac{1}{x}$ as we shall see in a later unit.

Try and get into the habit of writing this constant whenever you perform an integration – there doesn't seem to be much point at the moment but in the next section it becomes more important and we will need to calculate its precise value.

We can use this formula and two other simple rules to work out most simple algebraic integrals. The rules are:

1   Constants may be taken outside the integral, so for example:

$$\int 4x^8\,dx = 4\int x^8\,dx = 4 \times \tfrac{1}{9}x^9 + C$$
$$= \tfrac{4}{9}x^9 + C$$

2   The integral of a sum or difference is the same as the sum or difference of the integrals, so that:

$$\int (x^5 + x^7)\,dx = \int x^5\,dx + \int x^7\,dx$$
$$= \tfrac{1}{6}x^6 + \tfrac{1}{8}x^8 + C$$

You'll see that in the last example the two constants from the two integrals are combined into one constant – we don't need to write something like

$$\tfrac{1}{6}x^6 + C + \tfrac{1}{8}x^8 + D \text{ for instance.}$$

Note that a constant like 4 can be written $4x^0$ (since $x^0 = 1$), so

$$\int 4\,dx = 4\int x^0\,dx = 4 \times \frac{x^1}{1} + C = 4x + C$$

Example	Integrate the following functions:

(a) $\int (3x^2 - 5x + 7) \, dx$ 　　 (b) $\int \left(x^2 - \dfrac{1}{2x^2} \right) dx$ 　　 (c) $\int \left(6\sqrt{x} + \dfrac{1}{3\sqrt{x}} \right) dx$

Solution	(a)	To begin with, you might prefer to look at the integrals separately:

$$\int (3x^2 - 5x + 7) \, dx = \int 3x^2 \, dx - 5\int x \, dx + \int 7 \, dx$$

$$= 3 \times \frac{x^3}{3} - 5 \times \frac{x^2}{2} + 7x + C \quad = x^3 - \frac{5}{2} x^2 + 7x + C$$

After some practice you should be able to write them straight down.

(b) We need to express any x terms first of all in the form x^k to apply our formula. Careful with the constants: remember that $\dfrac{1}{2x^2}$ is *not* the same as $2x^{-2}$.

$$\int \left(x^2 - \frac{1}{2x^2} \right) dx \quad = \int \left(x^2 - \frac{1}{2} \times (x^{-2}) \right) dx$$

$$= \frac{x^3}{3} - \frac{1}{2} \times \frac{x^{-1}}{-1} + C \quad = \frac{x^3}{3} + \frac{1}{2x} + C$$

(c) $\int \left(6\sqrt{x} + \dfrac{1}{3\sqrt{x}} \right) dx \quad = \int \left(6x^{\frac{1}{2}} + \dfrac{1}{3} x^{-\frac{1}{2}} \right) dx$

$$= \frac{6x^{\frac{3}{2}}}{\frac{3}{2}} + \frac{1}{3} \frac{x^{\frac{1}{2}}}{\frac{1}{2}} + C \quad = 4 x^{\frac{3}{2}} + \frac{2}{3} x^{\frac{1}{2}} + C$$

[Remember that to divide by a fraction is the same thing as turning it upside–down and multiplying, so that

$$\int x^{\frac{3}{2}} \, dx \quad = \frac{x^{\frac{5}{2}}}{\frac{5}{2}} + C \quad = \frac{2}{5} x^{\frac{5}{2}} + C$$

Try and use fractions as much as possible, avoiding decimals.]

Practice questions A

1　Integrate the following functions, (remember + C):

(a) x^4 　　 (b) x 　　 (c) $\dfrac{1}{2}$ 　　 (d) x^9 　　 (e) $\dfrac{1}{x^2}$ 　　 (f) $\dfrac{2}{x^3}$

(g) $\dfrac{1}{5x^4}$ 　　 (h) $\dfrac{3}{4x^5}$ 　　 (i) $\sqrt{x}$ 　　 (j) $2x^{\frac{1}{3}}$ 　　 (k) $\dfrac{1}{4\sqrt{x}}$ 　　 (l) $x^{\frac{4}{5}}$

2　Find:

(a) $\int (5x^4 - 6x^2 + 8) \, dx$ 　　 (b) $\int \left(\dfrac{3}{x^2} + \dfrac{4}{x^3} \right) dx$ 　　 (c) $\int \left(2\sqrt{x} - \dfrac{6}{\sqrt{x}} \right) dx$

(d) $\int (x^2 - x - 1) \, dx$ 　　 (e) $\int (5x^3 - 2x^2 + 7x - 9) \, dx$ 　　 (f) $\int (10x^{\frac{3}{2}} - 8x^{\frac{1}{3}}) \, dx$

3   Multiply the brackets out first, and then find:

(a) $\int x\,(x-3)\ \mathrm{d}x$ (b) $\int x^2\,(1+x)\ \mathrm{d}x$ (c) $\int x^{\frac{1}{2}}\,(x+1)\ \mathrm{d}x$ (d) $\int x^{-2}\,(4+x^2)\ \mathrm{d}x$

(e) $\int (x^2-1)\,(x^2+1)\ \mathrm{d}x$ (f) $\int \left(x+\dfrac{1}{x}\right)^2\ \mathrm{d}x$ (g) $\int (3x+1)\,(x-2)\ \mathrm{d}x$ (h) $\int (x^{\frac{1}{2}}+1)\,(x^{-\frac{1}{2}}-1)\ \mathrm{d}x$

Solving $\dfrac{\mathrm{d}y}{\mathrm{d}x} = \mathrm{f}(x)$

OCR P1 5.1.6 (b)

This is actually the first step along a very long road called *Differential Equations*. The solution of these equations can give us information about the state of different physical situations at any time: how fast heat will flow, what the air-resistance will be for a given shape, what would be a reasonable prediction for a population in two years time, etc, etc.

As we saw in the last section, when we differentiate we lose information, since the original constant term (if any) disappears. To replace this loss, we need to know a specific additional piece of information. Taking a simple example, suppose we are told only that $\dfrac{\mathrm{d}y}{\mathrm{d}x} = 1$. We can see that if we differentiate any of

$y = x + 3, y = x + 1, y = x - 1$ etc, we have $\dfrac{\mathrm{d}y}{\mathrm{d}x} = 1$. So without any other

information, we can only say that the solution is $y = x + c$ which represents an (infinite) number of lines with a gradient of 1.

Figure 8.1

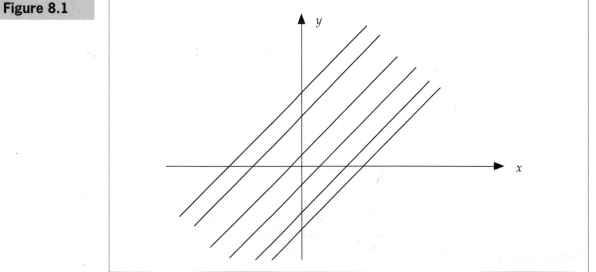

This is called the *general solution*. However, if we are told that the line passes through (1, 3), we can feed this into the general solution to give us a value for c: $3 = 1 + c \Rightarrow c = 2$. Our general solution is now the *particular* solution $y = x + 2$.

Figure 8.2

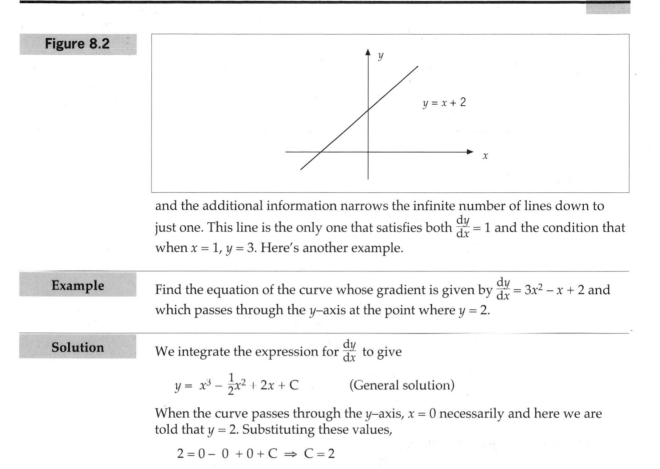

and the additional information narrows the infinite number of lines down to just one. This line is the only one that satisfies both $\frac{dy}{dx} = 1$ and the condition that when $x = 1$, $y = 3$. Here's another example.

Example

Find the equation of the curve whose gradient is given by $\frac{dy}{dx} = 3x^2 - x + 2$ and which passes through the y–axis at the point where $y = 2$.

Solution

We integrate the expression for $\frac{dy}{dx}$ to give

$$y = x^3 - \frac{1}{2}x^2 + 2x + C \qquad \text{(General solution)}$$

When the curve passes through the y–axis, $x = 0$ necessarily and here we are told that $y = 2$. Substituting these values,

$$2 = 0 - 0 + 0 + C \Rightarrow C = 2$$

Equation of the curve is $y = x^3 - \frac{1}{2}x^2 + 2x + 2$ (Particular solution)

Practice questions B

1 Find the general solution for the curves whose gradients are given by:

(a) $\frac{dy}{dx} = x^2 - 1$

(b) $\frac{dy}{dx} = \sqrt{x}$

(c) $\frac{dy}{dx} = \frac{1}{x^2} - \frac{1}{x^3}$

(d) $\frac{dy}{dx} = 6x^2 + 7x + 3$

2 Find the particular solution for the curves whose gradients are given by:

(a) $\frac{dy}{dx} = 2x - 1$, passing through $(1, 1)$

(b) $\frac{dy}{dx} = \sqrt{x} + \frac{1}{\sqrt{x}}$, passing through $(4, 2)$

(c) $\frac{dy}{dx} = x(x^2 + 4)$, passing through $(2, 2)$

(d) $\frac{dy}{dx} = x^{-1}(x^3 + x)$, passing through $(6, 2)$

(e) $\frac{dy}{dx} = 5x^3 + 6x^2 - 8x + 9$, passing through $(0, 5)$

(f) $\frac{dy}{dx} = 2x^2 + 4$, crossing the y–axis where $y = 4$

Definite integrals

Integration can give us the area under curves whose defining function we can integrate. For example, suppose we wanted to find the area under the curve $y = x^2$ bounded by the curve, the x-axis and the lines $x = 1$ and $x = 2$:

Figure 8.3

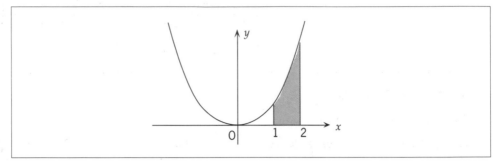

We would write this as $\int_1^2 x^2 dx$, the top limit being the larger of the two values.

When we've found the integral of the function we enclose it in square brackets with the limits at the top and bottom:

$$\int_1^2 x^2 \, dx \quad = \left[\frac{x^3}{3}\right]_1^2$$

Then, to evaluate this, we put in the top value instead of x, $\frac{2^3}{3} = \frac{8}{3}$, and from this subtract the value given by the bottom value instead of x, i.e. $\frac{1^3}{3} = \frac{1}{3}$

All in all,

$$\int_1^2 x^2 \, dx = \left[\frac{x^3}{3}\right]_1^2 = \frac{8}{3} - \frac{1}{3} = \frac{7}{3}$$

which is the area of the shaded region.

Note that when we're using definite integrals the constants are not included because they always disappear in the subtraction. Have a look at another example of this:

Example

Find the area underneath the curve of $y = x^3 - x$ between the points $x = 1$ and $x = 3$.

Solution

The curve of $y = x^3 - x$ looks something like this:

Figure 8.4

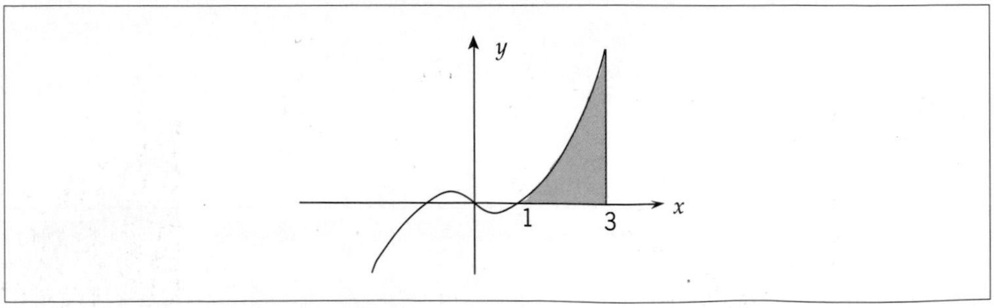

and the shaded portion is the area we want.

This is given by:

$$\int_1^3 y\,dx = \int_1^3 (x^3 - x)\,dx = \left[\frac{x^4}{4} - \frac{x^2}{2}\right]_1^3$$

$$= \left(\frac{3^4}{4} - \frac{3^2}{2}\right) - \left(\frac{1^4}{4} - \frac{1^2}{2}\right) = \left(\frac{81}{4} - \frac{9}{2}\right) - \left(\frac{1}{4} - \frac{1}{2}\right)$$

$$= \frac{63}{4} + \frac{1}{4} = 16$$

Practice questions C

1 Evaluate the following integrals:

(a) $\displaystyle\int_1^2 (4x + 1)\,dx$ (b) $\displaystyle\int_0^1 x^4\,dx$ (c) $\displaystyle\int_0^1 (1 - x^2)\,dx$ (d) $\displaystyle\int_1^4 \sqrt{x}\,dx$

(e) $\displaystyle\int_1^2 \frac{1}{x^2}\,dx$ (f) $\displaystyle\int_4^9 \frac{1}{\sqrt{x}}\,dx$ (g) $\displaystyle\int_1^2 (3x^2 - 4x + 3)\,dx$ (h) $\displaystyle\int_2^3 \frac{4}{5x^2}\,dx$

(i) $\displaystyle\int_0^4 x^{\frac{1}{2}}(x + 1)\,dx$ (j) $\displaystyle\int_1^2 x^{-2}(x^2 + 1)\,dx$ (k) $\displaystyle\int_{-2}^0 (x^3 - 4x)\,dx$ (l) $\displaystyle\int_0^1 (2 + x)(1 - x)\,dx$

2 Find the shaded areas:

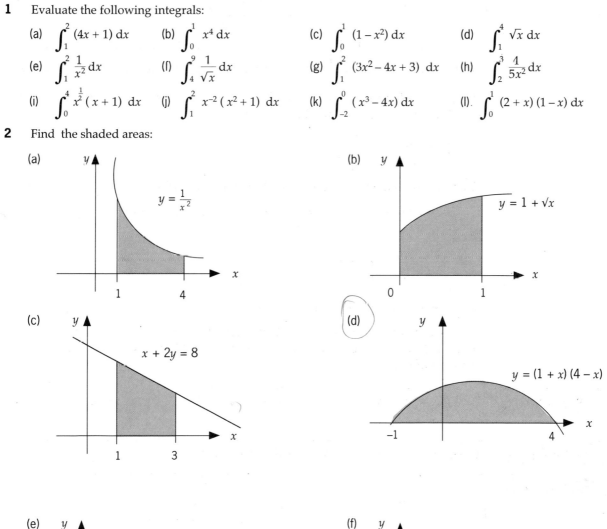

(a) $y = \dfrac{1}{x^2}$

(b) $y = 1 + \sqrt{x}$

(c) $x + 2y = 8$

(d) $y = (1 + x)(4 - x)$

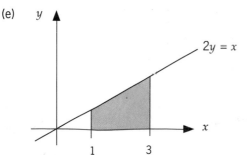

(e) $2y = x$

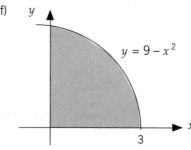

(f) $y = 9 - x^2$

3   Find the areas enclosed between the given curve, the x–axis and the given lines:

(a) $y = x^2$ $x = 1$ and $x = 3$ (b) $y = \sqrt{x}$ $x = 4$ and $x = 9$

(c) $y = \dfrac{2}{x^3}$ $x = 1$ and $x = 2$ (d) $y = 4 - \dfrac{2}{\sqrt{x}}$ $x = 1$ and $x = 4$

(e) $y = 2x^2 - x^3$ $x = \dfrac{1}{2}$ and $x = \dfrac{3}{2}$ (f) $y = 1 + \dfrac{1}{3x^2}$ $x = 3$ and $x = 6$

(g) $y = 2x^2 + 5$ $x = -1$ and $x = 2$ (h) $y = x^3 - 4x$ $x = -\dfrac{3}{4}$ and $x = -\dfrac{1}{2}$

Finding the limits

OCR P1 5.1.6 (d)

We may find that we are asked to find an area given only the equation of the curve and one of the boundary lines. In this case our first job is to solve the equation of the curve and the equation of the line simultaneously. This gives us the points of intersection and from this we can then proceed to integrate and calculate the area.

| **Example** | Find the area enclosed above the x–axis and below the curve given by $y = 4 - x^2$. |

| **Solution** | It helps to make a rough sketch first of all. |

Figure 8.5

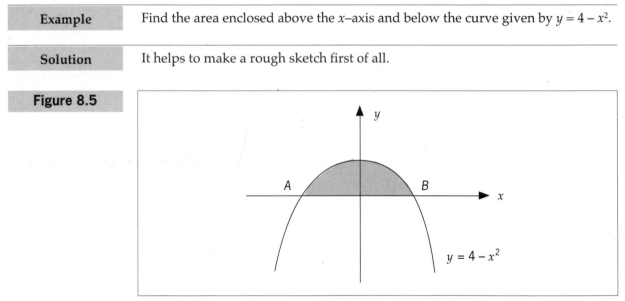

$y = 4 - x^2$

The area we need has been shaded. We need to find the coordinates of A and B these are at the intersection of the curve $y = 4 - x^2$ and the line $y = 0$ (i.e. the x–axis). Solving these simultaneously,

$$4 - x^2 = 0 \quad \Rightarrow \quad x^2 = 4 \quad \Rightarrow \quad x = \pm 2$$

We can now find the shaded area using these limits: it is just

$$\int_{-2}^{2} (4 - x^2)\, dx = \left[4x - \frac{x^3}{3} \right]_{-2}^{2}$$

$$= \left(8 - \frac{8}{3} \right) - \left(-8 - \left(-\frac{8}{3} \right) \right) = \frac{32}{3}$$

You have to be very careful with the signs in a case like this – there are a lot of minus signs around.

We can also be asked to find the areas enclosed between a curve and a line at an angle to the x–axis. We still have to find where the two intersect, but then we find the required area by finding two separate areas and subtracting.

Example

Find the points of intersection of the curve $y = x^2 + 2$ with the line $y = 3x$. Hence find the area enclosed between these two.

Solution

We draw a sketch first of all:

Figure 8.6

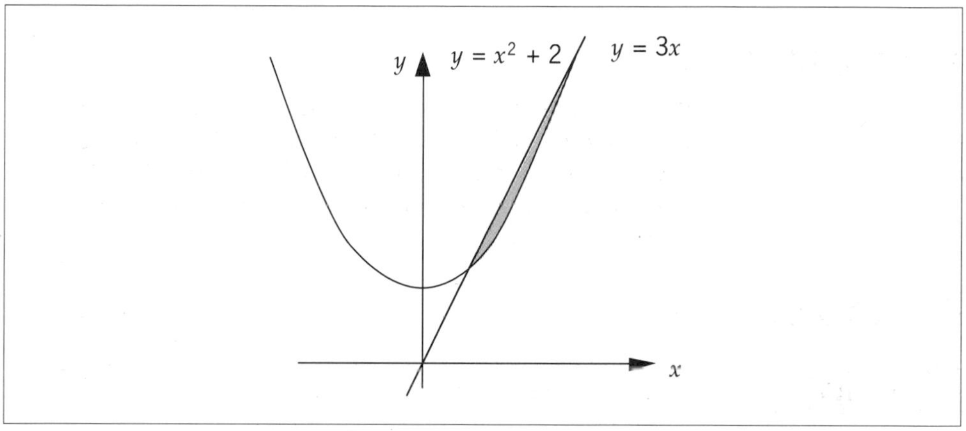

(y, $y = x^2 + 2$, $y = 3x$, x)

The curves intersect when $x^2 + 2 = 3x$

$$\Rightarrow x^2 - 3x + 2 = 0$$

$$(x - 2)(x - 1) = 0$$

The area shaded in the sketch is found by finding the area underneath the line, which is:

$$\int_1^2 3x \, dx = P, \text{ say}$$

and subtracting the area underneath the curve,

$$\int_1^2 (x^2 + 2) \, dx = Q, \text{ say}$$

Figure 8.7

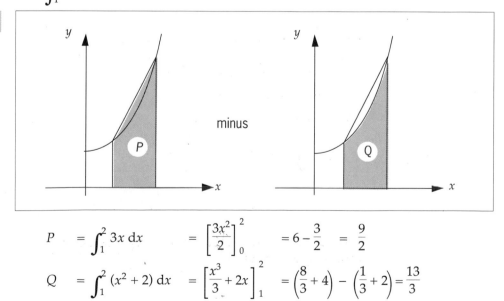

$$P = \int_1^2 3x \, dx = \left[\frac{3x^2}{2}\right]_0^2 = 6 - \frac{3}{2} = \frac{9}{2}$$

$$Q = \int_1^2 (x^2 + 2) \, dx = \left[\frac{x^3}{3} + 2x\right]_1^2 = \left(\frac{8}{3} + 4\right) - \left(\frac{1}{3} + 2\right) = \frac{13}{3}$$

The shaded region is then $P - Q = \dfrac{9}{2} - \dfrac{13}{3}$

$$= \frac{27 - 26}{6} = \frac{1}{6}$$

Practice questions D

1   Find the finite area enclosed by the x–axis and the curves:

  (a) $y = 1 - x^2$ (b) $y = (1 + x)(2 - x)$ (c) $y = 4x - x^2$ (d) $y = 6 + x - x^2$

2   Sketch the following pairs of lines and curves on the same axis. Find their points of intersection by solving the equations simultaneously and hence find the area enclosed between them

  (a) $y = x$ and $y = x^2$ (b) $y = x + 6$ and $y = x^2$

  (c) $y = 1 - x$ and $y = 1 - x^2$ (d) $y = 4 - x^2$ and $y = 1 - \dfrac{x}{2}$

Area under the x–axis

OCR P1 5.1.6 (d)

Any area of a curve which lies under the x–axis will work out to be *negative*.

| **Example** | Find the finite area between the curve $y = x^2 - 1$ and the x–axis. |

| **Solution** | |

| **Figure 8.8** | |

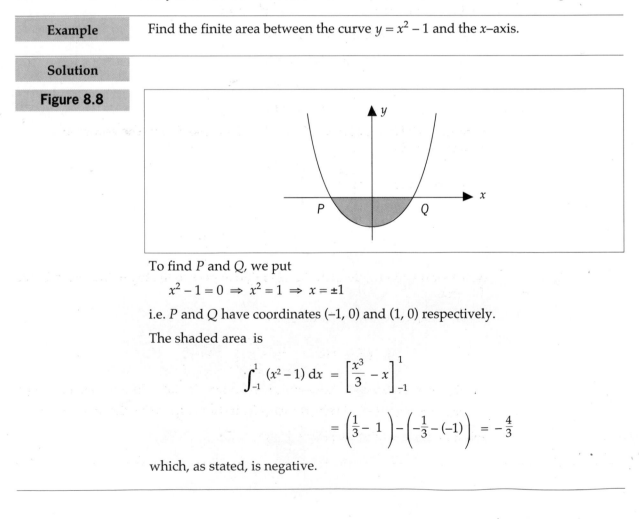

To find P and Q, we put

$$x^2 - 1 = 0 \implies x^2 = 1 \implies x = \pm 1$$

i.e. P and Q have coordinates $(-1, 0)$ and $(1, 0)$ respectively.

The shaded area is

$$\int_{-1}^{1} (x^2 - 1)\, dx = \left[\frac{x^3}{3} - x\right]_{-1}^{1}$$

$$= \left(\frac{1}{3} - 1\right) - \left(-\frac{1}{3} - (-1)\right) = -\frac{4}{3}$$

which, as stated, is negative.

If all of the area we need is under the axis there isn't a problem: we just ignore the minus sign. When part of the area is above and part below we have to be careful. We draw a sketch, find the different areas separately and then add the positive amounts together for a final answer.

Example

Find the finite area enclosed between the curve $y = x^3 - x$ and the x–axis.

Solution

A sketch first of all:

Figure 8.9

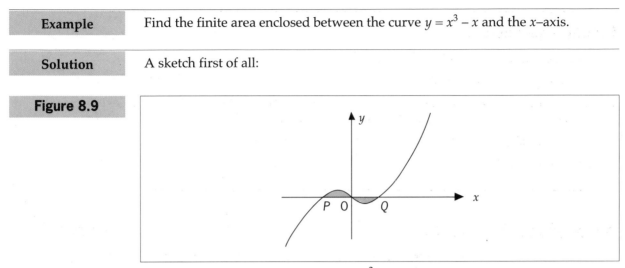

To find the points P and Q, we solve $x^3 - x = 0$

$$x^3 - x = 0 \quad \Rightarrow \quad x(x^2 - 1) = 0$$
$$\Rightarrow \quad x = 0 \text{ or } x^2 - 1 = 0$$
$$x^2 = 1$$
$$\rightarrow \quad x = \pm 1$$

So the x–coordinates of P and Q are –1 and 1 respectively. The shaded area between P and Q is

$$\int_{-1}^{0} (x^3 - x)\, dx \quad = \quad \left[\frac{x^4}{4} - \frac{x^2}{2} \right]_{-1}^{0}$$

$$= \quad 0 - \left(\frac{1}{4} - \frac{1}{2} \right) \quad = \quad \frac{1}{4}$$

Note that it's always the right-hand limit that goes to the top of the integral, 0 in this case.

The shaded area between 0 and Q is

$$\int_{0}^{1} (x^3 - x)\, dx \quad = \quad \left[\frac{x^4}{4} - \frac{x^2}{2} \right]_{0}^{1} \quad = \quad -\frac{1}{4}$$

which, since it's beneath the x–axis , is negative. We take the amount, ignoring the negative sign (technically, the *magnitude*) of $\frac{1}{4}$, add to this the other amount, and find that the total area is $\frac{1}{4} + \frac{1}{4} = \frac{1}{2}$

Practice questions E

1   Find the finite area enclosed between the given curve and the *x*–axis:

(a) $y = x^2 - 2x$

(b) $y = x^2 - 4$

(c) $y = (x - 1)(x - 3)$

(d) $y = (x + 2)(x + 4)$

2   Sketch the given curve and find the area enclosed between the curve, the *x*–axis and the given lines:

(a) $y = x^2 - 1$ between $x = 0$ and $x = 2$

(b) $y = 2 - x^2$ between $x = \dfrac{1}{\sqrt{2}}$ and $x = 2$

(c) $y = (3 + x)(1 - x)$ between $x = 0$ and $x = 2$

(d) $y = 2x - x^2$ between $x = -1$ and $x = 1$

Area between a curve and the y–axis

We saw that the formula for the area underneath a curve bounded by the two lines $x = x_2$ and $x = x_1$ and the *x*-axis is:

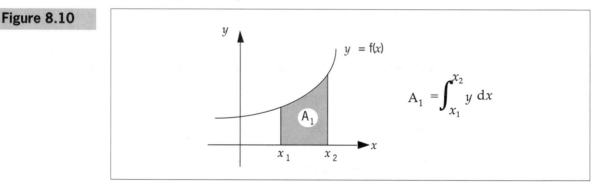

$$A_1 = \int_{x_1}^{x_2} y \, dx$$

If we wanted the area of the curve between the two lines $y = y_2$, $y = y_1$ and the *y*-axis, we need to swap everything round:

Figure 8.11

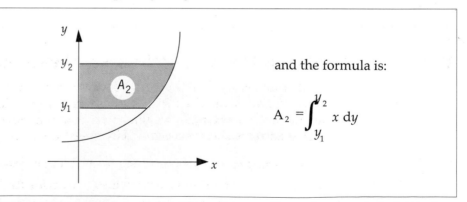

and the formula is:

$$A_2 = \int_{y_1}^{y_2} x \, dy$$

Note that if you are integrating with respect to *y* (that's the meaning of d*y*), the limits have to be the limits of *y*. We may need to change the original equation if this is given with *y* as the subject into one where *x* is the subject.

Let's have a look at an example of this.

Example	Find the area enclosed by the curve $y = x^2$, the lines $y = 4$, $y = 1$ and the y–axis.

Solution	

Figure 8.12	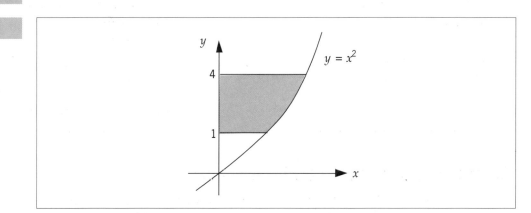

The formula for the area between the curve and the y–axis is

$$\int_{y_1}^{y_2} x \, \mathrm{d}y$$

so we need first of all to find out what x is in terms of y.

$$y = x^2 \quad \Rightarrow \quad y^{\frac{1}{2}} = x$$

The integral is now

$$\int_1^4 y^{\frac{1}{2}} \, \mathrm{d}y \quad = \quad \left[\frac{2}{3} y^{\frac{3}{2}} \right]_1^4$$

$$= \quad \frac{2}{3} \left(4^{\frac{3}{2}} - 1^{\frac{3}{2}} \right) = \frac{14}{3}$$

Definite integration as the limit of the sum of rectangles

In this section we shall have a quick look at the way in which we can approximate an area by the sum of a series of rectangles. By decreasing the width of these rectangles we can get some idea of how their sum approaches the exact value that we could find by integration.

As an example, we'll take the particular area under the curve $y = \frac{1}{x}$ between the limits $x = 1$ and $x = 2$. As a preliminary, we divide the line between the x-coordinates to obtain subdivisions at $x = 1\frac{1}{4}$, $1\frac{1}{2}$ and $1\frac{3}{4}$. We then construct two sets of four rectangles, one set lying wholly above the curve and the other wholly beneath it. Consequently the area under the curve, A say, will be more than the sum of the areas of the first set and less than the sum of the areas of the second set.

Figure 8.13

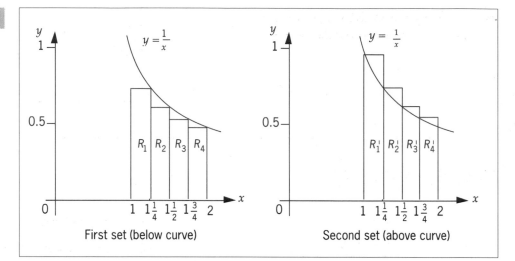

First set (below curve)   Second set (above curve)

We can work out that:

$$\sum_{i=1}^{4} = R_i = R_1 + R_2 + R_3 + R_4 = \frac{1}{5} + \frac{1}{6} + \frac{1}{7} + \frac{1}{8} \approx 0.63 \text{ (2 d.p.)}$$

and similarly:

$$\sum_{i=1}^{4} = R^1_i = \frac{1}{4} + \frac{1}{5} + \frac{1}{6} + \frac{1}{7} \approx 0.76 \text{ (2 d.p.)}$$

and so: $0.63 < A < 0.76$

We can increase the number of rectangles if we want to improve the accuracy: taking the number of strips to be n:

Table 8.1

Number of rectangles	Lower limit	Upper limit
$n = 10$	0.67	0.72
$n = 100$	0.691	0.696
$n = 1000$	0.6929	0.6934

The actual value of the area, found by integration, is 0.69315 (to 5 d.p.), which you can see is mid-way between the lower and upper limits.

Practice questions F

1 Find the areas enclosed by the given curves, the y–axis and the given line(s):

(a) $y = x + 1$ between $y = 2$ and $y = 4$

(b) $y = x^{\frac{1}{2}}$ between $y = 1$ and $y = 2$

(c) $y = 3x + 2$ between $y = 3$ and $y = 4$

(d) $y = 4x^2$ between $y = 1$ and $y = \frac{9}{4}$

(e) $y = \frac{1}{x^2}$ between $y = 1$ and $y = 4$

2 Find the area enclosed between the line $x + y = 6$, the y–axis and the line $y = 3$.

3 Find the area enclosed between the y–axis, the curve $y = x^2$ and the line $2x + y = 8$.

4 Find the finite area enclosed between the curve $y = (x - 1)^2$, the x–axis and the y–axis.

5   Find the area enclosed between the curve
$y = 1 + \sqrt{x}$, the y–axis and the line $y = 2$.

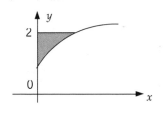

SUMMARY EXERCISE

1

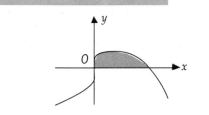

The diagram shows the graph of $y = \left(\sqrt[3]{x}\right) - x^2$.
Show by integration that the area of the region
(shaded in the diagram) between the curve and
the x-axis is $\frac{5}{12}$.

2   A curve for which

$$\frac{dy}{dx} = 1 + \frac{4}{x^2}$$

passes through the point $(-1, 4)$. Find:

(a)   the equation of the tangent to the curve
at the point $(-1, 4)$,

(b)   the equation of the curve.

3   A curve with equation $y = f(x)$ has

$$\frac{dy}{dx} = kx(3x + 2)$$

for each x, where k is a constant.

The curve passes through the point $(-1, 1)$
and has gradient 2 there.

(a)   Find the value of k.

(b)   Find an equation of the curve.   [AEB 1994]

4 $y = 3x^{\frac{1}{2}} - 4x^{-\frac{1}{2}}$, $x > 0$.

(a)   Find $\dfrac{dy}{dx}$.

(b)   Find $\displaystyle\int y\, dx$.

(c)   Hence show that $\displaystyle\int_{1}^{3} y\, dx = A + B\sqrt{3}$,

where A and B are integers to be found.

5   (a)   Find $\displaystyle\int \left(x^{\frac{1}{2}} - 4\right)\left(x^{-\frac{1}{2}} - 1\right) dx$.

(b)   Use your answer to part (a) to evaluate

$$\int_{1}^{3} \left(x^{\frac{1}{2}} - 4\right)\left(x^{-\frac{1}{2}} - 1\right) dx$$

giving your answer as an exact fraction.

6

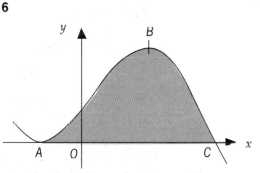

The above figure shows part of the curve
with equation $y = 3 + 5x + x^2 - x^3$. The curve
touches the x-axis at A and crosses the x-axis
at C. The points A and B are stationary
points on the curve.

(a)   Show that C has coordinates $(3, 0)$.

Using calculus and showing all your
working, find

(b)   the coordinates of A and B.

(c)   the area of shaded region, shown in the
figure, bounded by the curve and the x-
axis.

7

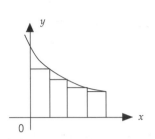

The diagram shows the part of the graph of

$$y = \frac{1}{1+x}$$

between $x = 0$ and $x = 1$. The four rectangles drawn under the curve are of equal width, and their total area is an approximation to the area under the curve from $x = 0$ to $x = 1$. Calculate this approximation to the area under the curve, giving 2 significant figures in your answer.

When there are n rectangles of equal width under the curve between $x = 0$ and $x = 1$ (instead of just 4), find an expression for their total area, and show that it may be written as

$$\sum_{r=1}^{n} \frac{1}{n+r}.$$

8

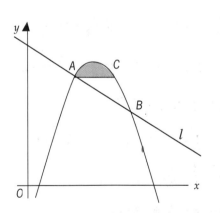

The above figure shows part of the curve with equation $y = p + 10x - x^2$, where p is a constant, and part of the line l with equation $y = qx + 25$, where q is a constant. The line l cuts the curve at the points A and B. The x-coordinates of A and B are 4 and 8 respectively. The line through A parallel to the x-axis intersects the curve again at the point C.

(a) Show that $p = -7$ and calculate the value of q.

(b) Calculate the coordinates of C.

The shaded region in the figure is bounded by the curve and the line AC.

(c) Using algebraic integration and showing all your working, calculate the area of the shaded region.

9

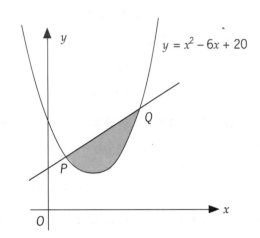

The above figure shows a sketch of the curve with equation

$$y = x^2 - 6x + 20$$

P is the point with the coordinates (2, 12).

(a) Show that the equation of the normal to the curve at P is $2y - x = 22$.

The normal to the curve at P meets the curve again at Q.

(b) Verify that the x-coordinate of Q is 4.5.

(c) Find the area of the finite region bounded by the line segment PQ and the arc QP of the curve, giving your answer to 3 significant figures.

10 The diagram shows part of the curve with equation $y = \frac{1}{x^2}$. Find the value of k for which area A has the same value as area B.

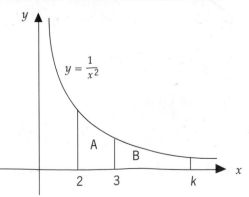

SUMMARY

When you have finished this section, you should:

- know that integration is the reverse process of differentiation
- know the general formula for integrating x^n
- remember the arbitrary constant, $+ C$, when you integrate
- know how to integrate algebraic expressions such as sums, differences, products and simple fractions
- know how to work out the general function of curves given their gradient function
- know how to find the particular function that passes through a given point
- appreciate that the area under a curve can be found using definite integration
- know how to find appropriate limits from consideration of the intersection of the curve with a given straight line
- know the formulae for finding the area between a curve and the (a) x-axis and (b) y-axis
- know that this formula gives a negative area when the curve lies under the x-axis
- in consequence of this, be prepared to split up a given area into parts and consider each separately
- know that the area under a curve can be approximated by a series of rectangles.

ANSWERS

Practice questions A

1 (a) $\frac{x^5}{5} + C$ (b) $\frac{x^2}{2} + C$ (c) $\frac{1}{2}x + C$

(d) $\frac{x^{10}}{10} + C$ (e) $\frac{-1}{x} + C$ (f) $\frac{-1}{x^2} + C$

(g) $-\frac{1}{15x^3} + C$ (h) $-\frac{3}{16x^4} + C$ (i) $\frac{2}{3}x^{\frac{3}{2}} + C$

(j) $\frac{3}{2}x^{\frac{4}{3}} + C$ (k) $\frac{\sqrt{x}}{2} + C$ (l) $\frac{5}{9}x^{\frac{9}{5}} + C$

2 (a) $x^5 - 2x^3 + 8x + C$ (b) $\frac{-3}{x} - \frac{2}{x^2} + C$

(c) $\frac{4}{3}x^{\frac{3}{2}} - 12\sqrt{x} + C$ (d) $\frac{x^3}{3} - \frac{x^2}{2} - x + C$

(e) $\frac{5x^4}{4} - \frac{2}{3}x^3 + \frac{7x^2}{2} - 9x + C$

(f) $4x^{\frac{5}{2}} - 6x^{\frac{4}{3}} + C$

3 (a) $\frac{x^3}{3} - \frac{3}{2}x^2 + C$ (b) $\frac{x^3}{3} + \frac{x^4}{4} + C$

(c) $\frac{2}{5}x^{\frac{5}{2}} + \frac{2}{3}x^{\frac{3}{2}} + C$ (d) $-\frac{4}{x} + x + C$

(e) $\frac{x^5}{5} - x + C$ (f) $\frac{x^3}{3} + 2x - \frac{1}{x} + C$

(g) $x^3 - \frac{5}{2}x^2 - 2x + C$ (h) $2x^{\frac{1}{2}} - \frac{2}{3}x^{\frac{3}{2}} + C$

Practice questions B

1 (a) $y = \frac{x^3}{3} - x + C$ (b) $y = \frac{2}{3}x^{\frac{3}{2}} + C$

(c) $y = \frac{-1}{x} + \frac{1}{2x^2}$

(d) $y = 2x^3 + \frac{7}{2}x^2 + 3x + C$

2 (a) $y = x^2 - x + 1$ (b) $y = \frac{2}{3}x^{\frac{3}{2}} + 2x^{\frac{1}{2}} - \frac{22}{3}$

(c) $y = \frac{x^4}{4} + 2x^2 - 10$

(d) $y = \frac{x^3}{3} + x - 76$

(e) $y = \frac{5}{4}x^4 + 2x^3 - 4x^2 + 9x + 5$

(f) $y = \frac{2}{3}x^3 + 4x + 4$

Practice questions C

1 (a) 7 (b) $\frac{1}{5}$ (c) $\frac{2}{3}$ (d) $\frac{14}{3}$

(e) $\frac{1}{2}$ (f) 2 (g) 4 (h) $\frac{2}{15}$

(i) $\frac{272}{15}$ (j) $\frac{3}{2}$ (k) 4 (l) $\frac{7}{6}$

2 (a) $\frac{3}{4}$ (b) $\frac{5}{3}$ (c) 6 (d) $\frac{125}{6}$

(e) 2 (f) 18

3 (a) $\frac{26}{3}$ (b) $\frac{38}{3}$ (c) $\frac{3}{4}$ (d) 8

(e) $\frac{11}{12}$ (f) $\frac{55}{18}$ (g) 21 (h) $\frac{11}{4}$

Practice questions D

1 (a) $\frac{4}{3}$ (b) $\frac{9}{2}$ (c) $\frac{32}{3}$ (d) $\frac{125}{6}$

2 (a)

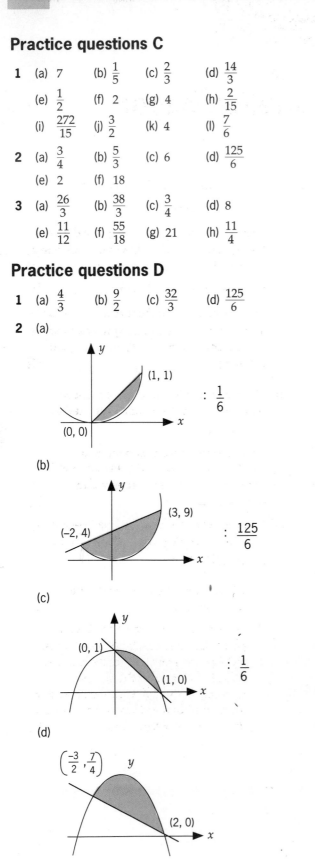

: $\frac{1}{6}$

(b)

: $\frac{125}{6}$

(c)

: $\frac{1}{6}$

(d)

Practice questions E

1 (a) $\frac{4}{3}$ (b) $\frac{32}{3}$ (c) $\frac{4}{3}$ (d) $\frac{4}{3}$

2 (a)

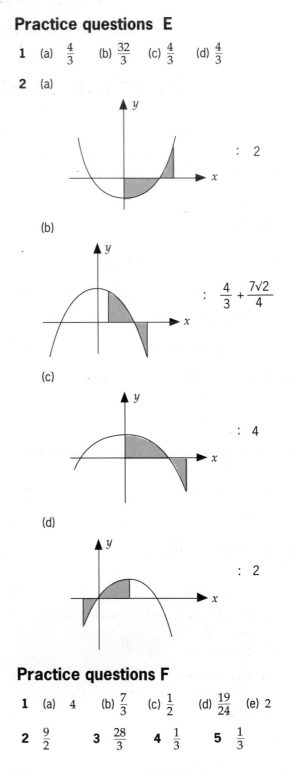

: 2

(b)

: $\frac{4}{3} + \frac{7\sqrt{2}}{4}$

(c)

: 4

(d)

: 2

Practice questions F

1 (a) 4 (b) $\frac{7}{3}$ (c) $\frac{1}{2}$ (d) $\frac{19}{24}$ (e) 2

2 $\frac{9}{2}$ 3 $\frac{28}{3}$ 4 $\frac{1}{3}$ 5 $\frac{1}{3}$

P1

Practice examination paper

Attempt all 9 questions. Available marks are given in brackets at the end of each question.

1 Find the set of values of x for which

$x(x-5) > 3(1-x)$ [7]

2 (a) Show that $\dfrac{x+1}{\sqrt{x}}$ can be written in the form

$x^a + x^b$, where a and b are constants.   [2]

(b) Evaluate $\displaystyle\int_4^9 \dfrac{x+1}{\sqrt{x}}\,dx$

giving your answer as a fraction.   [6]

3 Solve the simultaneous equations:

$x^2 - xy + y^2 = 7$
$2x - y = 5$ [8]

4 Find to the nearest $0.1°$ the values of x between $0°$ and $270°$ for which:

(a) $\sin 2x - 0.8$ [5]

(b) $4\tan^2 x = 1$ [4]

5 (a) Show that the equation

$2x^2 - 8x + 9 = 0$

has no real solutions.   [3]

(b) Find the values of the constants a, b and c for which

$2x^2 - 8x + 9 \equiv a(x+b)^2 + c$ [4]

(c) Sketch the curve of $y = 2x^2 - 8x + 9$ and state how this confirms the conclusion of part (a).   [4]

6 The curve for which $\dfrac{dy}{dx} = 3x^2 + ax + b$ has stationary points at $(-1, 5)$ and $(3, -27)$.

(a) Form two equations involving a and b using these points and solve these simultaneously to show that $a = -6$ and $b = -9$.   [5]

(b) Find the equation of the curve.   [5]

(c) Find the equation of the normal to the curve at the point where $x = 2$, giving your answer in the form $px + qy + r = 0$, where p, q and r are integers.   [6]

7 (a) The sum of the first n terms of a certain sequence is denoted S_n and is defined by

$S_n = n^2 + 4n$

Find and simplify an expression for $S_n - S_{n-1}$ and deduce that the series is arithmetic, giving the first term and common difference.   [4]

(b) (i) Show that the sum S_n of the first n terms of a geometric progression is given by

$S_n = \dfrac{a(1-r^n)}{1-r}$

where a is the first term and r the common ratio.   [5]

(ii) The amount of ore extracted from a particular mine falls by 10% each year. If the amount extracted in the first year was 360 tonnes, find the amount extracted in the fifth year and show that the total amount extracted in 9 years is a little more than 2200 tonnes.   [6]

8 The diagram shows a semicircle, centre O and radius r with AB a diameter. P lies on the circumference and N is the foot of the perpendicular from P onto AB. The angle PON is θ.

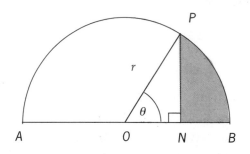

(a) Find the lengths of ON and PN in terms of r and θ, and deduce an expression for the area of $\triangle PON$. [5]

(b) Show that the area of the shaded region can be written:

$$A = \frac{1}{2}r^2(\theta - \sin\theta\cos\theta)$$ [3]

(c) If this area is $\frac{1}{4}$ of the area of the semi-circle, show that θ satisfies the equation:

$$\sin\theta\cos\theta = \theta - \frac{\pi}{4}$$ [4]

9 The diagram shows a triangle OAB with a rectangle $OPQR$ inside in such a way that Q lies on AB, $OA = 3$ and $OB = 4$, and the lengths of PQ and QR are x and y respectively.

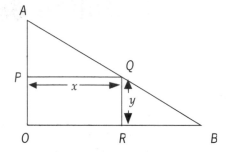

(a) Find the lengths of BR and AP in terms of x and y respectively. [2]

(b) Use similar triangles to show that $3x + 4y = 12$. [3]

(c) Deduce that the area A of the rectangle $OPQR$ can be written

$$A = 3x - \frac{3x^2}{4}$$ [3]

(d) Use differentiation to find the maximum area, showing that it is a maximum and not a minimum. [6]

Solutions

Section 1

1 $x^2 - 8x + 15 = 6 \implies (x - 4)^2 - 16 + 15 = 6$

$\implies (x - 4)^2 = 7$

$\implies x - 4 = \pm\sqrt{7}$ and $x = 4 + \sqrt{7}$ or $x = 4 - \sqrt{7}$

2 (a) $2(x^2 - 3x - 4) - (x^2 - 4x + 4) = 0$

$2x^2 - 6x - 8 - x^2 + 4x - 4 = 0$

$x^2 - 2x - 12 = 0$

$x = \dfrac{2 \pm \sqrt{4 + 48}}{2} = \dfrac{2 \pm 2\sqrt{13}}{2} = 1 \pm \sqrt{13}$

(b) $x > 1 + \sqrt{13}$ or $x < 1 - \sqrt{13}$

3 (a) $x^2 + x - 2 = 18$

$x^2 + x - 20 = 0$

$(x + 5)(x - 4) = 0 \implies x = -5$ or $x = 4$

(b) $x > 4$ or $x < -5$

4 $x^2 + 6x + 1 = kx^2 + k$

$(1 - k)x^2 + 6x + (1 - k)^2 = 0$

Equal roots $\implies b^2 - 4ac = 0$

$\implies 36 - 4(1 - k) = 0$

$4(1 - k)^2 = 36$

$(1 - k)^2 = 9$

$1 - k = \pm 3$

$\implies k = -2$ or 4

5 (a) $f(2) = 8 + 4 - 10 - 2 = 0 \implies (x - 2)$ is a factor

(b) $x^3 + x^2 - 5x - 2 = (x - 2)(x^2 + ax + 1)$

x^2–coeff: $1 = -2 + a \implies a = 3$

$f(x) = 0 \implies x = 2$ or $x^2 + 3x + 1 = 0, x = \dfrac{-3 \pm \sqrt{(9 - 4)}}{2}$

i.e. solutions are $2, \dfrac{-3 + \sqrt{5}}{2}$ and $\dfrac{-3 - \sqrt{5}}{2}$

6 $b^2 - 4ac > 0 \implies k^2 - 144 > 0 \implies k > 12$ or $k < -12$

7 (a) $3x^2 + 12x + 5 = 3[x^2 + 4x] + 5$

$= 3[(x + 2)^2 - 4] + 5$

$= 3(x + 2)^2 - 12 + 5$

$= 3(x + 2)^2 - 7$

$\implies p = 3, q = 2$ and $r = -7$

(b) Min value is -7

(c) $3(x + 2)^2 - 7 = 0 \implies 3(x + 2)^2 = 7$

$(x + 2)^2 = \dfrac{7}{3}$

$x + 2 = \pm\sqrt{\dfrac{7}{3}}$

$x = -2 \pm \sqrt{\dfrac{7}{3}}$

$\implies x = -0.5$ or -3.5 (1 dp)

8 $49x^2 - 21x + 2 = (7x - 1)(7x - 2)$

Equation is $(7\sqrt{y} - 1)(7\sqrt{y} - 2) = 0$

$\implies \sqrt{y} = \dfrac{1}{7} \implies y = \dfrac{1}{49}$

or $\sqrt{y} = \dfrac{2}{7} \implies y = \dfrac{4}{49}$

9 Area $= \dfrac{1}{2}ab \sin c = \dfrac{1}{2}x(x + 2)\sin 30° = 12$

$\implies x(x + 2) = 48 \ (\sin 30° = \dfrac{1}{2})$

$x^2 + 2x - 48 = 0 \quad (x + 8)(x - 6) = 0$

$\implies x = 6$ (since $x > 0$)

10 (a) $y = (x - p)^2 + p - p^2$

$\implies A$ has coords $(p, p - p^2)$

But $y > 0 \implies p - p^2 > 0 \quad p(1 - p) > 0$

$\implies 0 < p < 1$

(b) Substituting, $p - p^2 = 2p - 1$

$p^2 + p - 1 = 0$

$p = \dfrac{-1 \pm \sqrt{1 + 4}}{2} = \dfrac{-1 \pm \sqrt{5}}{2}$

But $p > 0 \implies p = \dfrac{-1 + \sqrt{5}}{2}$

11 (a) $f(-1) = 0 \implies -1 + a - b + 4 = 0 \qquad \ldots①$

$f(-2) = 0 \implies -8 + 4a - 2b + 4 = 0 \qquad \ldots②$

$① \times -2 \qquad 2 - 2a + 2b - 8 = 0 \qquad \ldots③$

$② + ③ \qquad -6 + 2a - 4 = 0$

$a = 5, b = 8$

(b) Put $x = 2 \implies 8 - 36 + 2A - 24 = 0 \implies A = 26$

$(x - 2)[x^2 - (B + C)x + BC] \equiv x^3 - 9x^2 + 26x - 24$

$x^2: \ -2 - (B + C) = -9 \implies B + C = 7$

157

$x = 0$ $-2BC = -24$ $\Rightarrow$ $BC = 12$

Inspection, $B = 4$ and $C = 3$ (or vice versa) and $A = 26$

(c) $f(x) = 4x^3 - 13x - 6$

$f(2) = 32 - 26 - 6 = 0$ $\Rightarrow x - 2$ is a factor

$(x - 2)(4x^2 + ax + 3) \equiv 4x^3 - 13x - 6$

$x^2 : -8 + a = 0 \Rightarrow a = 8$ $\Rightarrow (x - 2)(4x^2 + 8x + 3)$

$\Rightarrow$ $f(x) = (x - 2)(2x + 1)(2x + 3)$

12   (a)   Height Δ with BC as base is $3 - y$

$\Rightarrow \dfrac{3 - y}{x} = \dfrac{3}{4}$ $\left[\dfrac{\text{Height}}{\text{Base}}\right]$

$\Rightarrow$ $12 - 4y = 3x$ and $3x + 4y = 12$

(b) $S = xy = x\left[\dfrac{12 - 3x}{4}\right] = 3x - \dfrac{3x^2}{4}$

(c) $S = -\dfrac{3}{4}[x^2 - 4x] = -\dfrac{3}{4}\left[(x - 2)^2 - 4\right]$

$= 3 - \dfrac{3}{4}(x - 2)^2$

$a = 3, b = 2$

(d) $x = 2$ and $S = 3$

Section 2

1   As powers of 5, $\dfrac{(5^3)^x}{(5^2)^y} = 5^4$ $\Rightarrow \dfrac{5^{3x}}{5^{2y}} = 5$

$\Rightarrow 5^{3x - 2y} = 5^4$

i.e. $3x - 2y = 4$ ... ①

As power of 2, $2^1 \times (2^2)^x = (2^5)^y$

$\Rightarrow 2^1 \times 2^{2y} = 2^{5y}$ $\Rightarrow 2^{1 + 2x} = 2^{5y}$

i.e. $1 + 2x = 5y$ ...②

Solving simultaneously, $x = 2$ and $y = 1$

2   (a) $\dfrac{3}{\sqrt{2} - 1} - \dfrac{6}{\sqrt{2}} = \dfrac{3}{\sqrt{2} - 1} \times \dfrac{\sqrt{2} + 1}{\sqrt{2} + 1} - \dfrac{6}{\sqrt{2}} \times \dfrac{\sqrt{2}}{\sqrt{2}}$

$= \dfrac{3\sqrt{2} + 3}{2 - 1} - \dfrac{6\sqrt{2}}{2} = 3\sqrt{2} + 3 - 3\sqrt{2} = 3$

(b) $\dfrac{1}{3 - \sqrt{5}} + \dfrac{1}{3 + \sqrt{5}} = \dfrac{3 + \sqrt{5} + 3 - \sqrt{5}}{(3 - \sqrt{5})(3 + \sqrt{5})}$

$= \dfrac{6}{9 - 5} = \dfrac{6}{4} = \dfrac{3}{2}$

3   (a) $x^{\frac{2}{3}} = 8x \Rightarrow \dfrac{1}{8} = \dfrac{x}{x^{\frac{2}{3}}} \Rightarrow x^{\frac{1}{3}} = \dfrac{1}{8} \Rightarrow x = \dfrac{1}{512}$

(b) $2 \times (2^2)^{(x+1)} = (2^3)^x \Rightarrow 2^1 \times 2^{2x+2} = 2^{3x}$

$\Rightarrow 2x + 3 = 3x, \; x = 3$

4   (a) $\dfrac{-2}{3}$

(b) $\dfrac{1}{3\left(\frac{-1}{2}\right)^{-2}} = \dfrac{1}{3} \times \left(\dfrac{-1}{2}\right)^2 = \dfrac{1}{3} \times \dfrac{1}{4} = \dfrac{1}{12}$

5   If $y = 2^x \Rightarrow 2^{-x} = \dfrac{1}{2^x} = \dfrac{1}{y}$ and equation becomes

$2(y) + \dfrac{1}{y} = 3 \Rightarrow 2y^2 + 1 = 3y \equiv 2y^2 - 3y + 1 = 0$

$2y^2 - 3y + 1 = (2y - 1)(y - 1) = 0 \Rightarrow y = 1$ or $y = \dfrac{1}{2}$

$\Rightarrow 2^x = 1 \Rightarrow x = 0$ or $2^x = \dfrac{1}{2} \Rightarrow x = -1$

6   (a) $a^{p+q} = a^p \times a^q = 5 \times 9 = 45$

(b) $2a^{-p} = \dfrac{2}{a^p} = \dfrac{2}{5}$

(c) $a^{2p - \frac{1}{2}q} = a^{2p} \times a^{-\frac{1}{2}q} = (a^p)^2 \times (a^q)^{-\frac{1}{2}}$

$= 5^2 \times (9)^{-\frac{1}{2}} = 25 \times \dfrac{1}{3} = \dfrac{25}{3}$

7   If $y = x^{\frac{1}{3}}, \; x^{\frac{2}{3}} = \left(x^{\frac{1}{3}}\right)^2 = y^2$, equation becomes

$y^2 - 5y + 6 = 0 \Rightarrow (y - 3)(y - 2) = 0$

$\Rightarrow y = 3$ or $y = 2$ $\Rightarrow x^{\frac{1}{3}} = 3$ $\Rightarrow x = 27$

or $x^{\frac{1}{3}} = 2$ $\Rightarrow x = 8$

8 Using cosine-rule,

$(\sqrt{3})^2 = 3^2 + x^2 - 2 \times 3 \times x \times \cos 30°$

$\Rightarrow \quad 3 = 9 + x^2 - 3x\sqrt{3} \qquad \text{since } \cos 30° = \dfrac{\sqrt{3}}{2}$

$\Rightarrow \quad x^2 - (3\sqrt{3})\,x + 6 = 0$

Using quadratic formula, $x = \dfrac{3\sqrt{3} \pm \sqrt{(3\sqrt{3})^2 - 24}}{2}$

$\qquad\qquad = \dfrac{3\sqrt{3} \pm \sqrt{3}}{2}$

$\Rightarrow \quad x = \dfrac{3\sqrt{3} + \sqrt{3}}{2} = 2\sqrt{3} \ \text{ or } \ x = \dfrac{3\sqrt{3} - \sqrt{3}}{2} = \sqrt{3}$

9 (a) rearranging, $\quad x - 3 = \sqrt{x + 9}$

squaring $\qquad x^2 - 6x + 9 = x + 9$

collecting $\qquad x^2 - 7x = 0$

factorising $\qquad x\,(x - 7) = 0$

$\Rightarrow \quad x = 0 \ \text{ or } \ x = 7$

(b) squaring, $\qquad 4y - 9 = 4y - 4\sqrt{y} + 1$

rearranging, $\quad 4\sqrt{y} = 10$

$\sqrt{y} = \dfrac{5}{2} \qquad \Rightarrow \quad y = \dfrac{25}{4}$

10 (a) $5^2 = 25$ (b) $\dfrac{1}{4^{\frac{3}{2}}} = \dfrac{1}{\left(4^{\frac{1}{2}}\right)^3} = \dfrac{1}{8}$

(c) $\dfrac{8^{\frac{4}{3}}}{8^{\frac{2}{3}}} = 8^{\frac{2}{3}} = \left(8^{\frac{1}{3}}\right)^2 = 2^2 = 4$

11 If $y = 2^x$, $\quad 2^{x+2} - 2^x \times 2^2 = 4y$

$2^{x-1} = 2^x \times 2^{-1} = \dfrac{y}{2}$

Equation becomes $\quad 4y - 2 = 7 \times \dfrac{y}{2}$

$\dfrac{y}{2} = 2 \Rightarrow y = 4 \Rightarrow 2^x = 4 \Rightarrow x = 2$

12 (a) Using formula, $\quad x = \dfrac{-3\sqrt{3} \pm \sqrt{27 + 120}}{2}$

$= \dfrac{-3\sqrt{3} \pm \sqrt{147}}{2} = \dfrac{-3\sqrt{3} \pm 7\sqrt{3}}{2}$

$\Rightarrow \quad x = 2\sqrt{3} \ \text{ or } \ x = -5\sqrt{3}$

(b) Here $x = z^{\frac{1}{3}} \Rightarrow z^{\frac{1}{3}} = 2\sqrt{3} \Rightarrow z = 8 \times 3\sqrt{3} = 24\sqrt{3}$

or $z^{\frac{1}{3}} = -5\sqrt{3} \Rightarrow z = -125 \times 3\sqrt{3} = -375\sqrt{3}$

13 (a) (i) $b^{\frac{1}{8}} = \left(a^6\right)^{\frac{1}{8}} = a^{\frac{3}{4}}$

(ii) $c = \dfrac{a^{\frac{3}{4}} + 5\,a^{\frac{3}{4}}}{a^{\frac{1}{2}}} = \dfrac{6\,a^{\frac{3}{4}}}{a^{\frac{1}{2}}} = 6a^{\frac{1}{4}}$

and $k = 6,\ p = \dfrac{1}{4}$

(b) $\dfrac{16ab^2}{2a^{-1}b^2} = 8a^2$

14 $4X - Y = 1$

$X + Y - 1.5$

$\Rightarrow 5X = 2.5,\ X = \dfrac{1}{2},\ Y = 1$

$4^x = \dfrac{1}{2} \qquad \Rightarrow 2^{2x} = 2^{-1},\ x = -\dfrac{1}{2}$

$3^y = 1 \qquad \Rightarrow y = 0$

Section 3

1 (a) Area $= \frac{1}{2} r^2 \theta = \frac{1}{2} \times 5^2 \times 0.7 = 8.75 \text{ cm}^2$

(b) Area shaded = Area sector − Area Δ

$= 8.75 - \frac{1}{2} \times 5^2 \times \sin 0.7$

$= 0.70 \ (2 \text{ s.f.})$

2

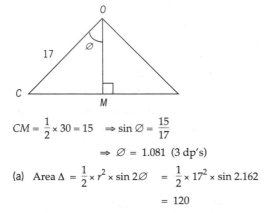

$CM = \frac{1}{2} \times 30 = 15 \quad \Rightarrow \sin \varnothing = \frac{15}{17}$

$\Rightarrow \varnothing = 1.081 \ (3 \text{ dp's})$

(a) Area $\Delta = \frac{1}{2} \times r^2 \times \sin 2\varnothing = \frac{1}{2} \times 17^2 \times \sin 2.162$

$= 120$

(b) $C\hat{O}D = 2.16 \ (2 \text{ dp's})$

(c) Area sector OCD bounded by A_2 is $\frac{1}{2} r^2 \theta$

$= \frac{1}{2} \times 17^2 \times 2.162 = 312.409$

$\Rightarrow$ Area curved segment CA_2D is $312.409 - 120$

$= 192.409$

Area semicircle on CD is $\frac{1}{2} \pi r^2 = 353.429$

$\Rightarrow$ Area R is $353.429 - 192.409 = 161.02 \text{ cm}^2$

3

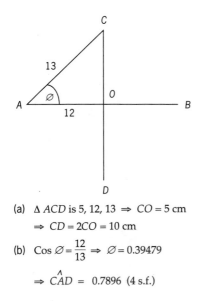

(a) ΔACD is 5, 12, 13 $\Rightarrow CO = 5$ cm

$\Rightarrow CD = 2CO = 10$ cm

(b) $\cos \varnothing = \frac{12}{13} \Rightarrow \varnothing = 0.39479$

$\Rightarrow C\hat{A}D = 0.7896 \ (4 \text{ s.f.})$

(c) Area sector CAD is $\frac{1}{2} r^2 \times C\hat{A}D = \frac{1}{2} \times 13^2 \times 0.7896$

$= 66.7197$

Area ΔCAD is $\frac{1}{2} r^2 \sin C\hat{A}D$

$= \frac{1}{2} \times 13^2 \times \sin (0.7896)$

$\Rightarrow$ Difference, $66.7197 - 60 = 6.7197$ is half shaded area, which is $13.4 \text{ cm}^2 \ (1 \text{ dp})$

4

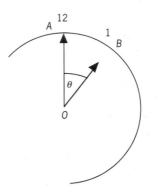

A travels the circumference of a circle centre O, radius 10 cm, i.e. 20π cm.

B travels $\frac{1}{12}$ circumference of a circle centre O, radius 6 cm, i.e. $\frac{1}{12} \times 2\pi \times 6 = \pi$ cm

Difference is $20\pi - \pi = 19\pi$.

5 $\frac{1}{2} x = 14.48° \ (\text{1st}) \text{ or } 165.52° \ (\text{2nd})$

$\Rightarrow x = 29.0 \text{ or } 331.0 \ (1 \text{ dp})$

6 $\cos^2 x = 1 - \sin^{2x} \Rightarrow 2(1 - \sin^2 x) + 3\sin x = 0$

$2\sin^2 x - 3\sin x - 2 = 0$

$(2\sin x + 1)(\sin x - 2) = 0$

$\Rightarrow \sin x = \frac{-1}{2} \Rightarrow x = 210° \text{ or } 330°$

or $\sin x = 2$ Impossible

7 $2\sin(3\theta - 48°) = 1 \qquad \Rightarrow \sin(3\theta - 48°) = \frac{1}{2}$

$3\theta - 48° = 30°, 150°, 390°$

$3\theta = 78°, 198°, 438° \qquad \Rightarrow \theta = 26°, 66°, 146°$

8 $\sin^2 3x = 1 - \cos^2 3x \Rightarrow 3(1 - \cos^2 3x) - 7\cos 3x - 5 = 0$

$3 - 3\cos^2 3x - 7\cos 3x - 5 = 0$

$3\cos^2 3x + 7\cos 3x + 2 = 0$

$(3\cos 3x + 1)(\cos 3x + 2) = 0$

$\cos 3x = \frac{-1}{3} \Rightarrow 3x = 109.5, \ 250.5, \ 469.5$

$\Rightarrow x = 36.5, 83.5, 156.5$ [no solution for $\cos 3x = -2$]

9 (a) $\theta + \dfrac{\pi}{3} = \dfrac{\pi}{4}$ or $\dfrac{5\pi}{4}$ or $\dfrac{9\pi}{4}$

$\Rightarrow \theta = \dfrac{-\pi}{12}$ or $\dfrac{11\pi}{12}$ or $\dfrac{23\pi}{12}$

In given interval, solutions are $\dfrac{11\pi}{12}$ or $\dfrac{23\pi}{12}$

(b) $2\theta = \pi + \dfrac{\pi}{3},\ 2\pi - \dfrac{\pi}{3},\ 3\pi + \dfrac{\pi}{3},\ 4\pi - \dfrac{\pi}{3}$

$\Rightarrow \theta = \dfrac{2\pi}{3},\ \dfrac{5\pi}{6},\ \dfrac{5\pi}{3},\ \dfrac{11\pi}{6}$

10 (a) (i) $\tan 15° = \dfrac{1}{\tan 75°} = \dfrac{1}{2+\sqrt{3}} \times \dfrac{2-\sqrt{3}}{2-\sqrt{3}}$

$= \dfrac{2-\sqrt{3}}{4-3} = 2-\sqrt{3}$

(ii) $\tan 105° = \tan(180-105°) = -\tan 75°$
$= -2-\sqrt{3}$

(b) $3\sin^2 x + \sin x - 2 = 0$

$(3\sin x - 2)(\sin x + 1) = 0$

$\sin x = \dfrac{2}{3} \Rightarrow x = 0.73$ or 2.41 (2 dp's)

or $\sin x = -1 \Rightarrow x = \dfrac{3\pi}{2} \Rightarrow 4.71$ (2 dp's)

11 A is in second quadrant (obtuse)

$\sin^2 A + \cos^2 A = 1 \Rightarrow \cos^2 A = 1 - \sin^2 A = 1 - \dfrac{1}{9} = \dfrac{8}{9}$

$\Rightarrow \cos A = \pm\sqrt{\dfrac{8}{9}} = \pm\dfrac{2\sqrt{2}}{3}$ $\cos < 0$ in 2nd,

so $\cos A = \dfrac{-2\sqrt{2}}{3}$.

12 (a) Max and min of cos are 1 and –1

$\Rightarrow$ greatest depth is $20 + 5 = 25$ m

least depth is $20 - 5 = 15$ m

(b)

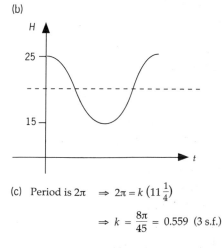

(c) Period is 2π $\Rightarrow 2\pi = k\left(11\frac{1}{4}\right)$

$\Rightarrow k = \dfrac{8\pi}{45} = 0.559$ (3 s.f.)

13 (a) $2\pi - \alpha,\ 2\pi + \alpha$

(b) $0 \le x < \alpha$ and $2\pi - \alpha < x < 2\pi + \alpha$

14 (a) Substituting $(15, 3 + \sqrt{3})$ gives

$3 + \sqrt{3} = 3 + 2\sin(20 + k)$

$\sqrt{3} = 2\sin(30 + k)$

$\sin(30 + k) = \dfrac{\sqrt{3}}{2}$

$30 + k = 60$ or 120

$\Rightarrow k = 30$ or 90

(b) $1 = 3 + 3\sin(2x + 30)$

$2\sin(2x + 30) = -2$

$\sin(2x + 30) = -1$

$2x + 30 = 270$ or 630

$2x = 240$ or $600 \Rightarrow x = 120$ or 300

(c) Max is $3 + 2 = 5$, Min is $3 - 2 = 1$

$\Rightarrow$ Range is $1 \le f(x) \le 5$

(d)

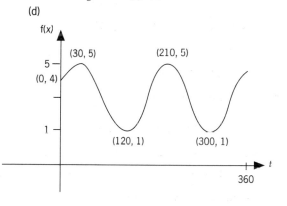

Section 4

1 (a) $\left(\dfrac{7-3}{2}, \dfrac{8+4}{2}\right)$ i.e. $(2, 6)$

   (b) Grad is $\dfrac{6+3}{2.-5} = -3 \Rightarrow y - 6 = -3(x - 2)$

     $y = -3x + 12$

2 (a) $6x + 4y = 2$ and $6x + 15y = 57$

     $\Rightarrow 11y = 55; \; y = 5, \; x = -3,$ i.e. $(-3, 5)$

   (b) Grad. is $\dfrac{-2}{5} \Rightarrow$ perp. grad. is $\dfrac{5}{2}, \; y - 3 = \dfrac{5}{2}(x - 7)$

     i.e. $2y = 5x - 29$

3 (a) $\sqrt{20^2 + 22^2} = \sqrt{884}$

   (b) Grad. is $\dfrac{22}{20} = \dfrac{11}{10} \Rightarrow y - 3 = \dfrac{11}{10}(x - 1)$

     $10y - 30 = 11x - 11 \Rightarrow 11x - 10y + 19 = 0$

4 (a) Grad. is $\dfrac{7}{-5} \Rightarrow y - 5 = \dfrac{7}{-5}(x + 1) \Rightarrow -5y + 25 = 7x + 7$

     i.e. $7x + 5y - 18 = 0$

   (b) $x = 0, \; y = \dfrac{18}{5} : y = 0, \; x = \dfrac{18}{7}$

     $\Rightarrow$ Area Δ is $\dfrac{1}{2} \times \dfrac{18}{7} \times \dfrac{18}{5} = 4.63$ (2 dp..)

5 (a) Grad. is $\dfrac{20}{-10} = -2 \Rightarrow y - 16 = -2(x - 2)$

               $\Rightarrow 2x + y = 20$

   (b) $y - 1 = \dfrac{1}{3}(x + 1) \Rightarrow 3y - 3 = x + 1, \; 3y = x + 4$

   (c) $x = 3y - 4 \Rightarrow 2(3y - 4) + y = 20.$

     $7y = 28 \Rightarrow y = 4, \; x = 8$

     D is $(8, 4) \Rightarrow OD^2 = 8^2 + 4^2 = 80$

     $\Rightarrow OD = \sqrt{80} = 4\sqrt{5}$

6 (a) $x = 0, \; y = 2; \; y = 0, \; x = \dfrac{-5}{6} \Rightarrow \dfrac{1}{2} \times \dfrac{5}{6} \times 2 = \dfrac{5}{6}$

   (b) Grad. l is $\dfrac{12}{5} \Rightarrow$ Grad. perp. is $\dfrac{-5}{12} \Rightarrow y = \dfrac{-5}{12}x$

   (c)

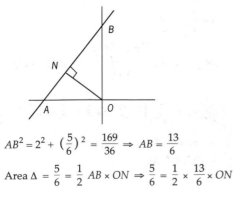

$AB^2 = 2^2 + \left(\dfrac{5}{6}\right)^2 = \dfrac{169}{36} \Rightarrow AB = \dfrac{13}{6}$

Area $\Delta = \dfrac{5}{6} = \dfrac{1}{2} AB \times ON \Rightarrow \dfrac{5}{6} = \dfrac{1}{2} \times \dfrac{13}{6} \times ON$

$\Rightarrow ON = \dfrac{10}{13}$, shortest distance

7 (a) Grad is $\dfrac{3}{2} \Rightarrow y = \dfrac{3}{2}x - \dfrac{3}{2}$

   (b) m is $y = 5 - \dfrac{2}{3}x \Rightarrow 5 - \dfrac{2}{3}x = \dfrac{3}{2}x - \dfrac{3}{2}$

     $\Rightarrow \dfrac{13}{6}x = \dfrac{13}{2}$

     $\Rightarrow x = 3, \; y = 3, \; C$ has coordinates $(3, 3)$

   (c) If P is on m and $x = -3, \; 2(-3) + 3y = 15 \Rightarrow y = 7$

     $PA^2 = (1 + 3)^2 + (0 - 7)^2 = 65$

     $PB^2 = (5 + 3^2)(6 - 7)^2 = 65$

     i.e. $PA = PB.$

8 (a) Grad. $\dfrac{2}{-4} = \dfrac{-1}{2} \Rightarrow y = \dfrac{-1}{2}x + 3$

   (b) $y - 0 = \dfrac{1}{4}(x + 9) \Rightarrow 4y = x + 9$

   (c) $\dfrac{1}{4}(x + 9) = \dfrac{-1}{2}x + 3 \Rightarrow \dfrac{3}{4}x = \dfrac{3}{4} \Rightarrow x = 1, \; y = \dfrac{5}{2}$

     $AD^2 = 1^2 + \left(\dfrac{1}{2}\right)^2 = \dfrac{5}{4} \Rightarrow AD = 1.12$ (2 d.p.)

   (d)

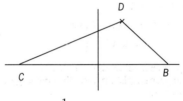

Area $\Delta = \dfrac{1}{2}$ base $\times$ height

       $= \dfrac{1}{2} \times 15 \times \dfrac{5}{2} = \dfrac{75}{4}$

9 (a) Grad. is $\dfrac{(3 + 4\sqrt{3}) - (3\sqrt{3})}{2 + \sqrt{3} - 1}$

     $= \dfrac{3 + \sqrt{3}}{1 + \sqrt{3}} \times \dfrac{1 - \sqrt{3}}{1 - \sqrt{3}} = \dfrac{3 + \sqrt{3} - 3\sqrt{3} - 3}{1 - 3}$

     $= \sqrt{3}$

   (b) $y - 3\sqrt{3} = \sqrt{3}(x - 1) = \sqrt{3}x - \sqrt{3} \Rightarrow y = \sqrt{3}x + 2\sqrt{3}$

   (c) $y = 0 \Rightarrow \sqrt{3}x + 2\sqrt{3} = 0 \Rightarrow x = -2 \Rightarrow C(-2, 0)$

   (d) $AC^2 = 3^2 + (3\sqrt{3})^2 = 36 \Rightarrow AC = 6$

   (e)

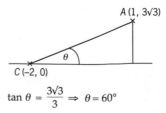

$\tan \theta = \dfrac{3\sqrt{3}}{3} \Rightarrow \theta = 60°$

10 Grad. $5x + y = 0$ is $-5 \Rightarrow$ equation of BD is

     $y - 2 = -5(x - 4)$ i.e. $5x + y = 22$

Grad. $5x - y = 0$ is 5

$\Rightarrow$ perp. grad. is $\frac{-1}{5}$

$\Rightarrow$ equation AC is $y - 5 = \frac{-1}{5}(x - 1)$

$\Rightarrow 5y - 25 = -x + 1, \ x + 5y = 26$

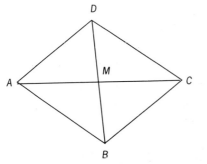

(a) $y = 22 - 5x \Rightarrow x + 5(22 - 5x) = 26$

$x + 110 - 25x = 26 \Rightarrow 84 = 24x, \ x = \frac{7}{2}, \ y = \frac{9}{2}$

(b) $A(1, 5) \rightarrow M\left(\frac{7}{2}, \frac{9}{2}\right) \rightarrow C(6, 4)$

$B(4, 2) \rightarrow M\left(\frac{7}{2}, \frac{9}{2}\right) \rightarrow D(3, 7)$

Grad. AD is $\frac{2}{2} = 1$

Grad $AB = -1 \Rightarrow$ perpendicular

Grad $BD = -5$, Grad $AC -\frac{1}{5}$

$\rightarrow$ not perpendicular, so since $ABCD$ parallellogram, right-angled, it's a rectangle (not a square)

Section 5

1 (a) $n = 1 \Rightarrow u_1 = 5 + (-1)^1 = 4$

$n = 2 \Rightarrow u_2 = 5 + (-1)^2 = 6$

$n = 3 \Rightarrow u_3 = 5 + (-1)^2 = 4$ and $u_4 = 6, u_5 = 4, u_6 = 6$

(b) $V_n = 9 + (-1)^n$

2 (a) $n = 1 \Rightarrow t_1 = \frac{1}{2} \times (2) = 1$

$t_2 = \frac{1}{2} \times 2(3) = 3 \Rightarrow t_1 + t_2 = 4$

$t_3 = \frac{1}{2} \times 3(4) = 6 \Rightarrow t_2 + t_3 = 9$

(b) $t_n + t_{n+1} = \frac{1}{2}n(n + 1) + \frac{1}{2}(n + 1)(n + 2)$

$= \frac{1}{2}(n + 1)[n + n + 2]$

$= \frac{1}{2}(n + 1)(2n + 2)$

$= (n + 1)(n + 1) = (n + 1)^2$

3 (a) $u_3 = 2 \times 3^2 = 18$

(b) $u_{n+1} - u_n = 2(n + 1)^2 - 2n^2$

$= 2[n^2 + 2n + 1 - n^2]$

$= 4n + 2$

(c) $a = u_2 - u_1 = 4(1) + 2 = 6$

$u_3 - u_2 = 4(2) + 2 = 10 \Rightarrow d = 4$

$a = 6, d = 4 \Rightarrow S_{1000} = \frac{1000}{2}[2 \times 6 + 999 \times 4]$

$= 2\,004\,000$

4 (a) $u_1 = 3\left(\frac{2}{3}\right) - 1 = 1 \qquad v_2 = 3\left(\frac{2}{3}\right)^2 - 1 = \frac{1}{3}$

$u_3 = 3\left(\frac{2}{3}\right)^2 - 1 = -\frac{1}{9}$

(b) $\displaystyle\sum_{n=1}^{15} 3\left(\frac{2}{3}\right)^n$

$= 3\left[\left(\frac{2}{3}\right) + \left(\frac{2}{3}\right)^2 + \ldots + \left(\frac{2}{3}\right)^{15}\right]$ GP,

$a = \frac{2}{3}, \ r = \frac{2}{3}$

$3S_{15} = 3 \times \dfrac{\frac{2}{3}\left(1 - \left(\frac{2}{3}\right)^{15}\right)}{1 - \frac{2}{3}} = 5.9863 \ (5 \ \text{s.f.})$

$\displaystyle\sum_{n=1}^{15} u_n = \sum_{n=1}^{15}\left[3\left(\frac{2}{3}\right)^n - 1\right] = \sum_{n=1}^{15} 3\left(\frac{2}{3}\right)^n - \sum_{n=1}^{15} 1$

$= 5.9863 - 15$

$= -9.014 \ (4 \ \text{sf})$

(c) $3u_{n+1}$ $= 3\left[3\left(\frac{2}{3}\right)^{n+1} - 1\right]$

$= 3\left[3 \times \frac{2}{3} \times \left(\frac{2}{3}\right)^n - 1\right]$

$= 3\left[2 \times \left(\frac{2}{3}\right)^n - 1\right]$

$= 2 \times 3 \times \left(\frac{2}{3}\right)^n - 3$

$= 2 \times 3\left(\frac{2}{3}\right)^n - 2 - 1$

$= 2\left[3\left(\frac{2}{3}\right)^n - 1\right] - 1$

$= 2u_n - 1$

5 (a) $u_1 = a = 3$ $u_7 = 2u_3 \Rightarrow a + 6d = 2(a + 2d)$
$= 2a + 4d \Rightarrow a = 2d$

$d = \frac{a}{2} = \frac{3}{2}$

(b) $S_{20} = \frac{20}{2}\left[6 + 19 \times \frac{3}{2}\right] = 345$

6 (a) $a + 8d = 7$
$a + 28d = 2(a + 4d) = 2a + 8d \Rightarrow a = 20d$
$20d + 8d = 7 \Rightarrow d = \frac{1}{4}, a = 5$

(b) $S_{200} = \frac{200}{2}\left[10 + 199 \times \frac{1}{4}\right] = 5975$

7 $S_{10} = 675$ $u_{10} = 2a$

Using $S_{10} = \frac{10}{2}[\text{First} + \text{Last}] = \frac{10}{2}[a + 2a] = 15a$
$\Rightarrow 15a = 675 \Rightarrow a = 45$ $a + 9d = 2a \Rightarrow a = 9d$
and $d = \frac{a}{9} = 5$

8 (a) £100 + £100 × 1.05 + £100 × 1.05^2 = £315.25
(b) GP $n = 40$, $a = 100$, $r = 1.05$

$S_{40} = \frac{100\left(1.05^{40} - 1\right)}{1.05 - 1}$ = £12079.98

9 $\frac{x-8}{2x} = \frac{2x+5}{x-8}$ $\Rightarrow$ $(x-8)^2 = 2x(2x+5)$

$x^2 - 16x + 64 = 4x^2 + 10x$

$3x^2 + 26x - 64 = 0$

$(3x + 32)(x - 2) = 0$

$\Rightarrow x = \frac{-32}{3}$ or $x = 2$

10 (a) $3n$

(b) GP, $a = -0.9$ $r = -0.9$ $S_n = \frac{-0.9\left(1 - (-0.9)^n\right)}{1 - (-0.9)}$

$= \frac{-0.9}{1.9}\left(1 - (-0.9)^n\right)$

(c) $u_1 = 0$ $u_2 = 2$ $u_3 = 0$ $u_4 = 2$
$\Rightarrow S_n = n$ (n even)
$= n - 1$ (n odd)

(d) AP, $a = 7$, $d = 3$, $S_n = \frac{n}{2}[14 + (n-1)3]$

$= \frac{n}{2}(3n + 11)$

11 (a) $u_3 = ar^2 = 27$, $u_6 = ar^5 = 8$ $\Rightarrow r^3 = \frac{8}{27} \Rightarrow r = \frac{2}{3}$

(b) $a \times \left(\frac{2}{3}\right)^2 = 27 \Rightarrow a = \frac{243}{4}$

(c) $S_\infty = \frac{a}{1-r} = \frac{\frac{243}{4}}{1 - \frac{2}{3}} = \frac{729}{4}$

(d) $S_{10} = \frac{243}{4} \frac{\left(1 - \left(\frac{2}{3}\right)^{10}\right)}{1 - \frac{2}{3}}$

$\Rightarrow$ difference is $\frac{729}{4}\left(\frac{2}{3}\right)^{10} = 3.16$ (3 s.f.)

12 (a) $\frac{15}{12} + \frac{1}{12} \times 1.2 = 1.35$ hrs $= 1$ hr 21 mins

(b) $u_{16} = \frac{1}{12} \times 1.2$ $u_{17} = \frac{1}{12} \times 1.2^2$ etc.

$u_r = \frac{1}{12} \times (1.2)^{r-15}$ for $16 \le r \le 25$

(c) $n = 10$ $S_{10} = \frac{0.1(1.2^{10} - 1)}{1.2 - 1} = 2.596$

$a = 0.1$
$r = 1.2$
$1.25 + 2.596 = 3$ hrs 51 mins

Section 7

1 (a) $y = x + 2x^{-1} \Rightarrow \dfrac{dy}{dx} = 1 - 2x^{-2}$

   (b) $\dfrac{dy}{dx} = \dfrac{-2}{x^3} \Rightarrow$ when $x = 2$, $\dfrac{dy}{dx} = \dfrac{-1}{4}$

2 $\dfrac{dy}{dx} = 2x - 6$, when $x = 2$ $\dfrac{dy}{dx} = -2$

   $\Rightarrow$ tangent is $y - 7 = -2(x - 4) \Rightarrow 2x + y = 15$

     is gradient at q $\dfrac{-1}{-2} = \dfrac{1}{2}$

   If $\dfrac{dy}{dx} = \dfrac{1}{2} \Rightarrow 2x - 6 = \dfrac{1}{2} \Rightarrow x = \dfrac{13}{4}$

3 If $x = 2$, $y = 4 \Rightarrow 4 = 4a + 2b$ ...①

   $\dfrac{dy}{dx} = 2ax + b \Rightarrow$ when $x = 2$, $4a + b = -8$ ...②

   $b = 12$, $a = -5$

4 (a) $\dfrac{dy}{dx} = 4x - 5$ If $\dfrac{dy}{dx} = 0$, $x = \dfrac{5}{4}$

   (b) When $x = 2$, $y = 1$ and $\dfrac{dy}{dx} = 3 \Rightarrow$ equation
of the normal is

     $y - 1 = \dfrac{-1}{3}(x - 2) \Rightarrow 3y - 3 = -x + 2$, $x + 3y = 5$

5 (a) $\dfrac{dy}{dx} = 4 - \dfrac{1}{x^2}$: when $\dfrac{dy}{dx} = 0$, $x = \pm\dfrac{1}{2}$

     $\rightarrow$ turning points $\left(\dfrac{1}{2}, 4\right)$ and $\left(\dfrac{-1}{2}, -4\right)$

   (b) $x = 2$, $\dfrac{dy}{dx} = \dfrac{15}{4}$ and $y = \dfrac{17}{2} \Rightarrow y - \dfrac{17}{2} = \dfrac{15}{4}(x - 2)$

     $4y - 34 = 15x - 30 \Rightarrow 4y = 15x + 4$

6 (a) $f'(x) = 3x^2 - 12x + 13$

   (b) $f'(x) = 3\left[x^2 - 4x\right] + 13$

         $= 3\left[(x - 2)^2 - 4\right] + 13$

         $= 3(x - 2)^2 + 1$

     i.e. $a = 3$, $b = 2$ and $c = 1$

     Since $f'(x) \geq 1 > 0$, f is an increasing function

7 $\dfrac{dy}{dx} = 2x + \dfrac{2}{x^2}$ when $x = 1$, $y = -1$ and $\dfrac{dy}{dx} = 4$

   $\Rightarrow$ equation tangent is $y + 1 = 4(x - 1)$

             $\Rightarrow$ $y = 4x - 5$

   Simultaneous, $4x - 5 = x^2 - \dfrac{2}{x}$

                $4x^2 - 5x = x^3 - 2$

                $x^3 - 4x^2 + 5x - 2 = 0$

                $(x - 1)^2(x - 2) = 0$ $x = 2$, $y = 3$

   Intersects at point $(2, 3)$

8 $\dfrac{dC}{dv} = -\dfrac{16\,000}{v^2} + 2v$. When $\dfrac{dC}{dv} = 0$, $2v = \dfrac{16\,000}{v^2}$

      $\Rightarrow v^3 = 8000$, $v = 20$

   $\dfrac{d^2C}{dv^2} = \dfrac{32\,000}{v^3} + 2 > 0 \Rightarrow$ MIN

9 (a) $\dfrac{dv}{dx} = 0.051 - 0.0006x$

     When $\dfrac{dv}{dx} = 0$, $x = \dfrac{510}{6} = 85$

     $\Rightarrow v = 5.6$ (2 s.f.).

   (b) Width is $2x = 170$ cm

10 $R = PS = P(140 - 3P^2) = 140P - 3P^3$

    $\dfrac{dR}{dP} = 140 - 9P^2$

    When $\dfrac{dR}{dP} = 0$, $P^2 = \dfrac{140}{9}$

    $\Rightarrow S = 140 - 3 \times \dfrac{140}{9} = \dfrac{280}{3}$

    $R = PS = \sqrt{\dfrac{140}{9}} \times \dfrac{280}{3} = £368.11$ when $P = £3.94$

11 $V = x^2(20 - x) = 20x^2 - x^3$

    $\dfrac{dV}{dx} = 40x - 3x^2$.

    When $\dfrac{dV}{dx} = 0$, $x(40 - 3x) = 0$

           $\Rightarrow x = 0$ or $x = \dfrac{40}{3}$

    $\dfrac{d^2V}{dx^2} = 40 - 6x$.

    When $x = \dfrac{40}{3}$, $\dfrac{d^2V}{dx^2} = -40 < 0 \Rightarrow$ MAX

12 (a)

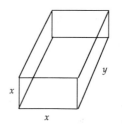

    Surface area of 5 faces is

    $2x^2 + 2xy + xy = 54$

    $\Rightarrow 2x^2 + 3xy = 54$ ... ①

    $V = x^2y$ ... ②

    From ①, $3xy = 54 - 2x^2 \Rightarrow y = \dfrac{54 - 2x^2}{3x}$

    Into ②, $V = x^2\left[\dfrac{54 - 2x^2}{3x}\right] = 18x - \dfrac{2}{3}x^3$

(b) $\dfrac{dV}{dx} = 18 - 2x^2$. When $\dfrac{dV}{dx} = 0$,

$2x^2 = 18 \Rightarrow x = \pm 3$ but $x > 0$

$\Rightarrow x = 3$

$V = 54 - \dfrac{2}{3}(27) = 36$

(c) $\dfrac{d^2V}{dx^2} = -4x < 0 \Rightarrow$ MAX

13 (a) If angle $MON = \theta$ rads, perimeter is $2r + r\theta = 100$

Area $= \dfrac{1}{2}r^2\theta = A$

From first equation, $r\theta = 100 - 2r \Rightarrow \theta = \dfrac{100}{r} - 2$

$\Rightarrow A = \dfrac{1}{2}r^2\left[\dfrac{100}{r} - 2\right] = 50r - r^2$

(b) $\dfrac{dA}{dr} = 50 - 2r = 0 \Rightarrow r = 25$

$\dfrac{d^2A}{dr^2} = -2 < 0 \Rightarrow$ MAX

(c) $\theta = \dfrac{100}{25} - 2 = 2$ (rads)

(d) $A = 50 \times 25 - 25^2 = 625 \text{ cm}^2$

14 (a) (i) Area $= 2rh + \dfrac{1}{2}\pi r^2 = 500$ ... ①

(ii) Perimeter $= 2r + 2h + \dfrac{1}{2}(2\pi r) = p$ ... ②

From ① $h = \dfrac{500 - \dfrac{1}{2}\pi r^2}{2r}$

Into ② gives $p = 2r + 2\left[\dfrac{500 - \dfrac{1}{2}\pi r^2}{2r}\right] + \pi r$

$= 2r + \dfrac{500}{r} - \dfrac{1}{2}\pi r + \pi r$

$= \left(2 + \dfrac{\pi}{2}\right)r + \dfrac{500}{r}$

(b) (i) $\dfrac{dp}{dr} = \left(2 + \dfrac{\pi}{2}\right) - \dfrac{500}{r^2}$

When $\dfrac{dp}{dr} = 0$, $\dfrac{500}{r^2} = 2 + \dfrac{\pi}{2} \Rightarrow r^2 = \dfrac{500}{2 + \dfrac{\pi}{2}}$

$\Rightarrow r = 11.8$ (3 s.f.)

(ii) $\dfrac{d^2p}{dr^2} = \dfrac{1000}{r^3} > 0 \Rightarrow$ MIN

15 (a)

Δ is isosceles, so $A\hat{D}E = 45°$ $(= D\hat{A}E)$

$AE = x \sin 45° = \dfrac{x}{\sqrt{2}} = DE$

Area Δ $\dfrac{1}{2}$base $\times$ height $= \dfrac{1}{2} \times \dfrac{x}{\sqrt{2}} \times \dfrac{x}{\sqrt{2}} = \dfrac{1}{4}x^2 \text{ m}^2$.

(b) Call the length EF l

$\Rightarrow$ Volume $=$ Area $\times l = \dfrac{1}{4}x^2 l = 4000$ ... ①

$S = 2 \times \dfrac{1}{4}x^2 + 2 \times \dfrac{x}{\sqrt{2}} \times l = \dfrac{1}{2}x^2 + \sqrt{2}x\,l$ ... ②

From ①, $l = \dfrac{16\,000}{x^2} \Rightarrow S = \dfrac{1}{2}x^2 + \sqrt{2}x \times \dfrac{16\,000}{x^2}$

$= \dfrac{x^2}{2} + \dfrac{16\,000\sqrt{2}}{x}$

(c) $\dfrac{dS}{dx} = x - \dfrac{16\,000\sqrt{2}}{x^2} = 0 \Rightarrow x^3 = 16\,000\sqrt{2}$

$\Rightarrow x^3 = 20\sqrt{2}$

$\Rightarrow S = 400 + 800 = 1200$

(d) $\dfrac{d^2S}{dx^2} = 1 + \dfrac{32\,000\sqrt{2}}{x^3} > 0 \Rightarrow$ MIN.

Section 8

1 $y = 0 \Rightarrow \sqrt[3]{x} - x^2 = 0 = x = 0$ or 1

$$\int_0^1 (x^{\frac{1}{3}} - x^2)\, dx = \left[\frac{3}{4}x^{\frac{4}{3}} - \frac{1}{3}x^3\right]_0^1$$

$$= \left(\frac{3}{4} - \frac{1}{3}\right) - 0 = \frac{5}{12}$$

2 (a) When $x = -1$, $\dfrac{dy}{dx} = 1 + 4 = 5$

$\Rightarrow$ equation $y - 4 = 5(x + 1)$

$\Rightarrow y = 5x + 9$

(b) $y = \int \left(1 + \dfrac{4}{x^2}\right) dx = x - \dfrac{4}{x} + C$

Substitute $(-1, 4)$

$4 = -1 + 4 + C \Rightarrow C = 1 \Rightarrow y = x - \dfrac{4}{x} + 1$

3 (a) When $x = -1$, $\dfrac{dy}{dx} = 2$

$\Rightarrow 2 = k(-1)(-3 + 2) \Rightarrow k = 2$

(b) $y = \int 2x(3x + 2)dx = 2x^3 + 2x^2 + C$

Substitute $(-1, 1)$

$1 = -2 + 2 + C \Rightarrow C = 1$

$\Rightarrow y = 2x^3 + 2x^2 + 1$

4 (a) $\dfrac{dy}{dx} = \dfrac{3}{2}x^{-\frac{1}{2}} + 2x^{-\frac{3}{2}}$

(b) $\int (3x^{\frac{1}{2}} - 4x^{-\frac{1}{2}})\, dx = 2x^{\frac{3}{2}} - 8x^{\frac{1}{2}} + C$

(c) $\left[2x^{\frac{3}{2}} - 8x^{\frac{1}{2}}\right]_1^3$

$= (2 \times 3\sqrt{3} - 8\sqrt{3}) - (2 - 8)$

$= 6 - 2\sqrt{3} \Rightarrow A = 6, B = -2$

5 (a) $I = \int (1 - x^{\frac{1}{2}} - 4x^{-\frac{1}{2}} + 4)\, dx = 5x - \dfrac{2}{3}x^{\frac{3}{2}} - 8x^{\frac{1}{2}} + C$

(b) $\left[5x - \dfrac{2}{3}x^{\frac{3}{2}} - 8x^{\frac{1}{2}}\right]_1^4 = \left(20 - \dfrac{16}{3} - 16\right) - \left(5 - \dfrac{2}{3} - 8\right)$

$= 7 - \dfrac{14}{3} = \dfrac{7}{3}$

6 (a) $f(x) = 3 + 5x + x^2 - x^3$

$f(3) = 0$ $\Rightarrow x - 3$ is a factor.

Since not repeated, C has coordinates $(3, 0)$

(b) $\dfrac{dy}{dx} = 5 + 2x - 3x^2 = (5 - 3x)(1 + x)$

When $\dfrac{dy}{dx} = 0$, $x = \dfrac{5}{3}$ or -1

$\Rightarrow$ coordinates of A are $(-1, 0)$, of B are $\left(\dfrac{5}{3}, \dfrac{256}{27}\right)$

(c) $\displaystyle\int_{-1}^{3} (3 + 5x + x^2 - x^3)\, dx$

$$= \left[3x + \frac{5}{2}x^2 + \frac{x^3}{3} - \frac{x^4}{4}\right]_{-1}^{3}$$

$$= \left(9 + \frac{45}{2} + 9 - \frac{81}{4}\right) - \left(-3 + \frac{5}{2} - \frac{1}{3} - \frac{1}{4}\right) = \frac{64}{3}$$

7 Area of rectangles is $\dfrac{1}{4}\left[\dfrac{4}{5} + \dfrac{4}{6} + \dfrac{4}{7} + \dfrac{4}{8}\right]$

$= 0.63$ (2 s.f.)

[Exact area is 0.69 (2 s.f.)]

Width is $\dfrac{1}{n}$ so area is $\dfrac{1}{n}\left[\dfrac{1}{1 + \dfrac{1}{n}} + \dfrac{1}{1 + \dfrac{2}{n}} + \dots + \dfrac{1}{1 + 1}\right]$

$= \dfrac{1}{n}\left[\dfrac{n}{n + 1} + \dfrac{n}{n + 2} + \dots + \dfrac{n}{2n}\right]$

$= \dfrac{1}{n + 1} + \dfrac{1}{n + 2} + \dots + \dfrac{1}{2n} = \displaystyle\sum_{r=1}^{n} \dfrac{1}{n + r}$

8 (a) Curve meets line when $p + 10x - x^2 = qx + 25$

When $x = 4$, at A, $p + 40 - 16 = 4q + 25$

$\Rightarrow p - 4q = 1$ ... ①

When $x = 8$, at B, $p + 80 - 64 = 8q + 25 \Rightarrow p - 8q = 9$ ... ②

① and ② simultaneously, $p = -7$ and $q = -2$

(b) Line through A that cuts at C is $y = 17$

$17 = -7 + 10x - x^2 \Rightarrow x^2 - 10x + 24 = 0 \Rightarrow x = 4$

(at A) or $x = 6$ (at C)

Coordinates of C are $(6, 17)$

(c) $\displaystyle\int_4^6 (-7 + 10x - x^2)dx = \left[-7x + 5x^2 - \dfrac{x^3}{3}\right]_4^6$

$= (-42 + 180 - 72) - \left(-28 + 80 - \dfrac{64}{3}\right) = \dfrac{106}{3}$

Subtract rectangle, $2 \times 17 = 34 \Rightarrow$ shaded area is

$\dfrac{106}{3} - 34 = \dfrac{4}{3}$

9 (a) $\dfrac{dy}{dx} = 2x - 6$ At p, $\dfrac{dy}{dx} = 4 - 6 = -2 \Rightarrow$

Grad. normal is $\dfrac{1}{2}$

Equation of the normal is $y - 12 = \dfrac{1}{2}(x - 2)$

$\Rightarrow 2y - 24 = x - 2$

$2y - x = 22$

(b) When $x = 4.5$, curve $y = \dfrac{81}{4} - 27 + 20 = \dfrac{53}{4}$

normal $2y = \dfrac{9}{2} + 22 = \dfrac{53}{2} \Rightarrow y = \dfrac{53}{4}$

i.e. intersect when $x = 4.5$

(c) Area under line PQ is trapezium,

$\frac{1}{2}(12 + \frac{53}{4}) \times \frac{5}{2} = 31.563$

Area under curve PQ is $\int_{2}^{9/2} (x^2 - 6x + 20)\,dx$

$= \left[\frac{x^3}{3} - 3x^2 + 20x\right]_{2}^{9/2}$

$= 28.958$

$\Rightarrow$ shaded area is 2.60 (3 s.f.)

10 $\int_{3}^{k} \frac{1}{x^2}\,dx = \int_{2}^{3} \frac{1}{x^2}\,dx$

$\Rightarrow \left[\frac{1}{-x}\right]_{3}^{k} = \left[\frac{1}{-x}\right]_{2}^{3}$

$\Rightarrow \frac{-1}{k} + \frac{1}{3} = \frac{-1}{3} + \frac{1}{2}$

$\frac{1}{k} = \frac{2}{3} - \frac{1}{2} = \frac{1}{6}$

$\Rightarrow k = 6$